Christmas

at the Royal Institution

Christmas

at the Royal Institution

An Anthology of Lectures by **M Faraday, J Tyndall, R S Ball, S P Thompson,**
E R Lankester, W H Bragg, W L Bragg, R L Gregory, and **I Stewart**

Editor

Frank A J L James

The Royal Institution of Great Britain, London

World Scientific

NEW JERSEY · LONDON · SINGAPORE · BEIJING · SHANGHAI · HONG KONG · TAIPEI · CHENNAI

Published by

World Scientific Publishing Co. Pte. Ltd.

5 Toh Tuck Link, Singapore 596224

USA office: 27 Warren Street, Suite 401-402, Hackensack, NJ 07601

UK office: 57 Shelton Street, Covent Garden, London WC2H 9HE

British Library Cataloguing-in-Publication Data
A catalogue record for this book is available from the British Library.

Christmas at the Royal Institution: An Anthology of Lectures by M Faraday, J Tyndall, R S Ball, S P Thompson, E R Lankester, W H Bragg, W L Bragg, R L Gregory, and I Stewart

ISBN-13 978-981-277-108-7
ISBN-10 981-277-108-5
ISBN-13 978-981-277-109-4 (pbk)
ISBN-10 981-277-109-3 (pbk)

Typeset by Stallion Press
Email: enquiries@stallionpress.com

Printed in Singapore by B & JO Enterprise

Preface

I am delighted to write this preface to this collection of Royal Institution Christmas Lectures, edited by Professor James. As Director of The Royal Institution (Ri), I am very proud that our annual flagship event is being commemorated in this way. The Christmas Lectures represent the core of what The Ri stands for: the engagement of young people through experiment-driven exploration. The lectures are unique in engaging the audience in both the design and interpretation of experiments — an approach all too sadly neglected these days in the dash to meet the targets of a dense scientific curriculum.

For the last 50 years, the Christmas Lectures have reached out to an ever-wider audience comprising a large majority of adults through the now well-known broadcast over the Christmas period. These broadcasts have been, for some time, a familiar part of the traditional landscape of the British Christmas holidays. As such, they empower the general public to appreciate not just cutting-edge scientific discoveries, but also the joy and excitement of asking a question that can be tested empirically. Anyone who has watched these broadcasts will know how different they are from the standard scientific programmes: far from relying on extensive emphasis on outside broadcasting, the Christmas Lectures instead adopt a much more modest yet realistic approach, often using everyday objects and situations familiar to young people. As the youngest ever Nobel Prize winner and Director of The Ri, Lawrence Bragg, remarked, "Never talk about science, *show* it to them". If anything, this is the mission statement of The Royal Institution in general and the Christmas Lectures in particular.

As Professor James shows in the forthcoming pages, this tradition has been seamless since Faraday's time until the present day. I myself was privileged to give the 1994 series; I therefore feel particularly familiar with the thrills and spills that inevitably occur. I can quite honestly claim that just before the beginning of the first lecture was one of the most frightening times of my life. Imagine standing in front of closed doors behind which waited some 400 children, several live animals, and five television cameras. As the monitor screen counted down the last minute, I finally understood the term "legs turning to jelly". Anyone who has watched the Christmas Lectures will know that invariably so many things can go wrong — after all, this is real science. In my own experience, a cockerel did not crow when it was supposed to, but did give full voice offstage 20 minutes later during a completely different demonstration. However, it is the ability of the lecturer to improvise with appropriate explanations when the results are unexpected that gives adults and children alike a true insight into scientific methods.

The Christmas Lectures then are unique and represent the very special agenda of The Royal Instruction: to diffuse science for the common purposes of life. In the spirit of Faraday, the blacksmith's son who went on to discover electromagnetic induction, the Christmas Lectures are truly democratising and have never been needed as much as they are today. This anthology stands testament not just to the narrative of the science of the past, but also to the importance of this approach for the science of the future.

Susan Greenfield
Director of the Royal Institution
Fullerian Professor of Physiology and Comparative Anatomy

Acknowledgements

The idea for this anthology of Christmas Lectures at The Royal Institution arose out of a conversation that I had with Dr Sonia Ojo, acquisitions editor at Imperial College Press and formerly a colleague of mine at The Royal Institution, at the Imperial College launch of *The Life and Scientific Legacy of George Porter*. I am thus enormously grateful to Dr Ojo for passing on the idea to World Scientific Publishing, the partners of Imperial College Press. At World Scientific in Singapore, Ms Wanda Tan has been the most efficient of editors and I gratefully acknowledge all of her support.

World Scientific and I wish to thank Professor Richard Gregory and Professor Ian Stewart for their permission to republish their lectures, as well as the Orion Publishing Group who originally published both lectures. We also thank The Royal Institution for permission to publish Figures 1 to 5 of the introduction, as well as, acting on behalf of the Bragg family, for permission to republish the lectures of William Henry Bragg and William Lawrence Bragg.

Finally, I thank my colleague at The Royal Institution, Ms Jane Harrison, for scanning all of the images from all 11 lectures published here.

Frank A. J. L. James

Contents

Introduction

Frank A. J. L. James

As Baroness Greenfield remarks in her Preface, thanks to television The Royal Institution's Christmas Lectures for young people are now as familiar a part of British Christmas as roast turkey or the Queen's Speech. In the lectures, eminent scientists illustrate the basics, and refer on occasion to some of the more advanced aspects, of their particular subjects using a wide variety of illustrations and experimental demonstrations. As Lawrence Bragg (1890–1971) once said to his lecture assistant, Bill Coates (1919–1993), "[N]ever talk about science, *show* it to them".[1] It is this visual, and sometimes aural and tactile, quality of the lectures that gives them such a special appeal, as Charles Taylor (1922–2002) — one of the most successful post-1945 demonstration lecturers — fully recognised.[2]

The Christmas Lectures have been televised annually in Britain since the 1966–1967 series delivered by the engineer Eric Laithwaite (1921–1997) (Fig. 1). Viewing figures regularly pass the one-million mark; indeed, so popular and well known have the lectures now

---

[1] Quoted in W. A. Coates, "Sir William [sic] Bragg and his lecturer's assistant," in John M. Thomas and David Phillips (eds.), *Selections and Reflections: The Legacy of Sir Lawrence Bragg*, Science Reviews, Northwood, 1990, pp. 147–9 on p. 149.
[2] Charles Taylor, *The Art and Science of Lecture Demonstration*, Adam Hilger (IOP Publishing), Bristol, 1988.

Fig. 1. Photograph of Eric Laithwaite delivering a Christmas Lecture during the
1966–1967 series, the first to be televised.

become that over the past decade they have been repeated and
broadcast in Japan, Korea, and most recently Brazil. But, their televising (and now webcasting) is only the current endpoint in an
effort, now nearly a century and a half old, to convey the content of
the lectures beyond those few hundred fortunate enough to be present in The Royal Institution's theatre for each lecture.

The Christmas Lectures evolved out of afternoon lectures which
had been delivered at The Royal Institution since 1800, the year
after its founding. The Royal Institution was founded at a meeting
held on 7 March 1799 in the Soho Square house of the President of
the Royal Society, Joseph Banks (1743–1820). Established during
the war against France, the original purpose of The Royal
Institution was to provide lectures and scientific advice in order to
help support and improve agriculture, industry, and the British

Empire. By June 1799, The Royal Institution had purchased the head lease on 21 Albemarle Street in Mayfair, where, almost uniquely for a scientific institution, it has remained ever since. The building first had to be converted from a gentleman's townhouse, with parts dating back to the early 18th century, into a scientific institution, with lecture theatres, laboratories, display areas, libraries, meeting rooms, and so on. The Corinthian columns of The Royal Institution that so dominate the northern end of Albemarle Street were not added until the 1830s (Fig. 2).[3]

The semicircular steeply raked lecture theatre on the first floor was designed by Thomas Webster (1773–1844) and construction

Fig. 2. Water colour by Thomas Hosmer Shepherd of the facade of The Royal Institution, erected in the late 1830s.

---

[3] For an account of the history of the building and how changes to it affected The Royal Institution's scientific work, see Frank A. J. L. James and Anthony Peers, "Constructing space for science at The Royal Institution of Great Britain," *Physics in Perspective* **9**: 130–85, 2007.

Fig. 3. Etching by James Gillray showing the administration of nitrous oxide (laughing gas) to Sir John Hippisley at a Royal Institution afternoon lecture in the early 19th century. Humphry Davy is holding the bellows.

was completed by 1802, when the caricaturist James Gillray (1757–1815) depicted a rather exciting afternoon lecture (Fig. 3). On occasion, this theatre could hold more than 1000 people; and it remained fundamentally unaltered until The Royal Institution's electric substation exploded on 29 December 1927, just after a Christmas Lecture had been delivered by Edward Neville da Costa Andrade (1887–1971). The thought of what might have happened had the explosion occurred while there were several hundred children in the building prompted the start of a major rebuilding programme. This involved the demolition of the old theatre and the construction, to the same design and layout, of a new theatre which could hold about 430 people.

This lecture theatre, in both its forms, has been and continues to be the place where The Royal Institution carries out one of its original missions of communicating science to a general audience. Shortly after its foundation, The Royal Institution was given a great boost in doing this when Humphry Davy (1778–1829) joined in 1801. His afternoon lectures were very popular and thereby made Albemarle Street the first one-way street in London, owing to the large numbers who flocked to see and hear him. Not only was Davy an immensely attractive lecturer, who firmly established The Royal Institution as *the* venue in London for spectacular and engaging lectures on science; but he also turned The Royal Institution into a place for scientific research, which had not been originally envisaged by the founders. He discovered sodium and potassium,[4] among other chemical elements, and invented a form of the miner's safety lamp.[5] He married a wealthy heiress in 1812 and was thus able to retire from being Professor of Chemistry at The Royal Institution at the age of 34, although he retained considerable influence there for the rest of his life. Thus, it was on Davy's recommendation in the spring of 1813 that Michael Faraday (1791–1867) was appointed as an assistant in the laboratory. Faraday had recently completed his apprenticeship as a bookbinder, but he was so interested in science that he applied to Davy for a job at The Royal Institution.[6]

Although Davy was succeeded as Professor of Chemistry at The Royal Institution by William Thomas Brande (1788–1866), it was really Faraday who inherited Davy's mantle as the most popular scientific lecturer in London.[7] Faraday was promoted through The

---

[4] David Knight, *Humphry Davy: Science and Power*, Blackwell, Oxford, 1992, pp. 65–9.

[5] Frank A. J. L. James, "How big is a hole? The problems of the practical application of science in the invention of the miners' safety lamp by Humphry Davy and George Stephenson in Late Regency England," *Transactions of the Newcomen Society* **75**: 175–227, 2005.

[6] For details, see Frank A. J. L. James, *The Correspondence of Michael Faraday*, Vol. 1, Institution of Electrical Engineers, London, 1991, pp. xxx–xxxi. Hereafter cited as Faraday, *Correspondence*, followed by volume, year, and letter number.

[7] Geoffrey Cantor, "Educating the judgment: Faraday as a lecturer," *Bulletin for the History of Chemistry* **11**: 28–36, 1991.

Royal Institution hierarchy and was appointed Director of the Laboratory in 1825. His outstanding research abilities, including his discoveries, amongst others, of electromagnetic rotations and induction (the principles behind the electric motor, transformer, and generator), led to the establishment of the Fullerian Professorship of Chemistry created especially for him in 1833.

Faraday did not deliver his first lecture at The Royal Institution until 1824, when he joined Brande in giving chemistry lecture courses for medical students, mostly from the nearby St George's Hospital. With the exception of a series of lectures delivered at the London Institution in 1827 (the same year that he began delivering afternoon lecture courses at The Royal Institution) and his lectures to cadets at the Royal Military Academy, Woolwich, from 1830 to 1852,[8] all of Faraday's lectures were delivered at The Royal Institution. In the mid-1820s, he initiated the Friday Evening Discourses, a series which — like the Christmas Lectures that began at about the same time — continues to this day.[9]

The Christmas Lectures were specifically tailored for what was then termed "juveniles", which meant those in the age range of 15–20, although this range probably went down in later years.[10] The original intention in 1825 was to provide "a set of Twenty two Lectures on Natural Philosophy suited to a Juvenile Auditory, during the Christmas, Easter, and Whitsuntide recesses".[11] It was

---

[8] Frank A. J. L. James, "The military context of chemistry: the case of Michael Faraday," *Bulletin for the History of Chemistry* **11**: 36–40, 1991.

[9] Frank A. J. L. James, "Running the Royal Institution: Faraday as an administrator," in Frank A. J. L. James (ed)., *The Common Purposes of Life: Science and Society at the Royal Institution of Great Britain*, Ashgate, Aldershot, 2002, pp. 119–46.

[10] Sophie Forgan, *The Royal Institution of Great Britain, 1840–1873*, unpublished PhD thesis, University of London, 1977, pp. 191–2.

[11] Minutes of the meeting of Royal Institution Managers, 5 December 1825, in the Archives of The Royal Institution, Vol. 7, p. 43. Advertised in *Times*, 17 December 1825, p. 1a.

agreed that John Millington (1779–1868), Professor of Mechanics at The Royal Institution, would deliver the first series. Very little is known about that series; and the second, given the following year by a now obscure but then popular astronomy lecturer named John Wallis,[12] is equally little documented. The Royal Institution ran a second Easter series in 1827, given by the Institution's Professor of Natural History, John Harwood (c. 1794–1854), but this was not a success; from then onwards, Juvenile Lectures were given only around Christmas.[13] Although they remained officially denominated as the Juvenile Lectures, it is clear from correspondence that certainly by the 1850s they were being familiarly termed as the Christmas Lectures and by the 1860s in print.[14]

The third series in 1827–1828, and the first by Faraday, was on the subject of chemistry. Thereafter, he dominated the Christmas Lectures until the early 1860s, delivering a further 18 series and in many ways establishing the format of the lectures, with each series usually comprising six lectures.[15] Brande delivered seven series between 1834–1835 and 1850–1851, when he retired from The Royal Institution, and Faraday gave all of the other series during the 1850s. Their successors, John Tyndall (1820–1893) and James Dewar (1842–1923), delivered 12 and 9 series, respectively. Combined with those series delivered by some of the other professors at The Royal Institution, such as Edward Frankland (1825–1899), William Odling (1829–1921), and John Hall Gladstone

---

[12] J. N. Hays, "The London lecturing empire, 1780–1850," in Ian Inkster and Jack Morrell (eds.), *Metropolis and Province: Science in British Culture, 1780–1850*, Hutchinson, London, 1983, pp. 91–119 on p. 99.

[13] For a detailed study of the early years of the Christmas Lectures, see Forgan, *op. cit.* (10), pp. 188–94.

[14] For example, Faraday to Schoenbein, 9 December 1852, Faraday, *Correspondence*, Vol. 4, 1999, letter 2604; "Michael Faraday," *Illustrated London News* **38**: 28–9 on p. 29, 1861.

[15] Faraday's lecture notes for most of his Christmas Lectures are in the archives of The Royal Institution.

(1827–1902), the professoriate dominated the Christmas Lecture programme until the 1890s, with comparatively few contributions from outside. From then onwards, the professors at The Royal Institution provided comparatively few of the Christmas Lectures, and this remains the case. Of course, some professors in the 20th century, such as William Bragg (1862–1942) and Lawrence Bragg, did deliver lectures; in their cases, four and two series each. The lecture on electrostatics in the second series delivered by Lawrence Bragg on "Electricity" (1961–1962) was depicted in a large oil painting by Terence Cuneo (1907–1996) (Fig. 4). This image, which well illustrates Bragg's commitment to showing science, was commissioned by The

Fig. 4. Oil painting by Terence Cuneo showing Lawrence Bragg delivering a Christmas Lecture on electrostatics in the 1961–1962 series. Bill Coates is operating the Wimshurst machine and the President of The Royal Institution, Lord Brabazon, is on the chair.

Royal Institution as a token of their gratitude for Bragg's role in sustaining them through a very difficult period.[16]

By the 1850s, the Christmas Lectures had become sufficiently well known for Faraday to be shown in an engraving published in the *Illustrated London News* delivering a lecture before Prince Albert (1819–1861), the Prince of Wales (1841–1910), and Prince Alfred (1844–1900), not paying as much attention as one might hope (Fig. 5) (a version of this image was used on the Bank of England £20 note depicting Faraday issued during the 1990s). By the 1860–1861 series, Faraday and the Christmas Lectures had become indissolubly linked

Fig. 5. Engraving by Alexander Blaikley published in *Illustrated London News* **28**: 177, 1856. It shows Michael Faraday delivering a Christmas Lecture on 27 December 1855 before Prince Albert, the Prince of Wales, and Prince Alfred (not paying too much attention).

---

[16] Frank A. J. L. James and Viviane Quirke, "L'affaire Andrade or how *not* to modernise a traditional institution," in James, *op. cit.* (9), pp. 273–304 on pp. 300–301.

in the public mind. In a profile of Faraday, the *Illustrated London News* in early 1861 commented, "For the last eight seasons Professor Faraday has undertaken this task with a modesty and a power which it is impossible to praise too much. There can be no greater treat to any one fond of scientific pursuits than to attend a course of these lectures".[17]

Publishers, realising the popularity of Faraday's Christmas Lectures, sought to persuade him to allow them to be published, reportedly offering him almost unlimited terms for the copyright.[18] Until the 1860s, Faraday declined all such invitations, being convinced that it was impossible to turn live lectures, with a large number of experimental demonstrations, into print form. Even as late as 1859, he wrote to one publisher about his rejection of an earlier offer: "[It was] proposed to take them by short hand & so save me trouble — but I knew that would be a thorough failure. Even if I cared to give time to the revision of the M.S. still the Lectures without the experiments & the vivacity of speaking would fall far behind those in the lecture room".[19]

Faraday's change of mind may be linked to his opposition to the rise of spiritualism, especially table turning, in the early 1850s. Many linked these phenomena to electricity and magnetism, and Faraday was deluged with requests for an explanation in these terms. After attending a couple of séances, he concluded that table turning was due to an involuntary muscular action on the part of the participants. He reported this in articles in the *Times* and *Athenaeum*,[20] which resulted in his being sent even more letters

---

[17] "Michael Faraday," *Illustrated London News* **38**: 28–9 on p. 29, 1861.
[18] *Ibid.*
[19] Faraday to Smith, 3 January 1859, in Faraday, *Correspondence*, Vol. 5, 2007, letter 3541.
[20] Faraday to the Editor of the *Times*, 28 June 1853, in Faraday, *Correspondence*, Vol. 4, 1999, letter 2691; "Professor Faraday on table-moving," *Athenaeum*, 2 July 1853, pp. 801–3.

attesting to the reality of the phenomena. Faraday was appalled at the existence of such scientific ignorance, which also offended his religious beliefs,[21] in an otherwise well-educated country. Faraday's response was to help arrange in 1854 a special course of lectures at The Royal Institution delivered by a number of eminent savants to demonstrate the value of scientific education; his contribution was a lecture on mental education.[22] Having set out his agenda, he pursued it actively over the next decade or so in a number of ways, including finally agreeing to have his Christmas Lectures published.

The key agent in the publishing process was, ironically in view of his later strong commitment to psychical research, the young chemist and journalist, William Crookes (1832–1919), whom Faraday seems to have first met at one of the séances.[23] In late 1859, Crookes had founded the weekly *Chemical News* and, always in need of copy, persuaded Faraday to allow one of his staff to take shorthand notes of his lectures in the 1859–1860 series on "The Various Forces of Matter". These were published in the *Chemical News*,[24] where each lecture occupied roughly a quarter of an issue. These were then collected together in book form, the first series of Christmas Lectures to be published, and the final

---

[21] Geoffrey Cantor, *Michael Faraday: Sandemanian and Scientist: A Study of Science and Religion in the Nineteenth Century*, Macmillan, London, 1991, pp. 148–50.

[22] Michael Faraday, "Observations on mental education," in *Lectures on Education Delivered at the Royal Institution of Great Britain*, The Royal Institution, London, 1854.

[23] William Crookes, *Psychic Force and Modern Spiritualism: A Reply to the "Quarterly Review" and Other Critics*, Longman, London, 1871, p. 11.

[24] Michael Faraday, "A course of six lectures (adapted to a juvenile auditory), consisting of illustrations of the various forces of matter, i.e. of such as are called the physical or inorganic forces — including an account of their relations to each other," *Chemical News* **1**: 52–5, 65–8, 77–80, 88–91, 100–3, 126–9, 1860.

lecture is reprinted here. Crookes had managed to solve the problem of how to turn an illustrated live lecture, with experimental demonstrations, into printed form for a popular market. That Faraday was satisfied with the outcome is confirmed by the fact that the following year the procedure was repeated for his lectures on the "Chemical History of a Candle" (in which he recycled his notes from his 1848–1849 series)[25] and it was published as a book in April 1861. It would appear that Faraday had transferred gratis his copyright in both books to Crookes, who used the sales income to support the difficult finances of the *Chemical News*.[26]

The Chemical History of a Candle must count as one of the most successful popular science books ever published. It has hardly been out of print since the 1860s and has been translated into many languages including French, Polish, Japanese, and Basic English.[27] It has thus been used as an example of good practice in a study on science communication.[28] Furthermore, the structure of the title has been imitated by others. For instance, George Porter (1920–2002), Director of The Royal Institution from 1966 to 1986, entitled his 1976–1977 Christmas Lectures "The Natural History of a Sunbeam"; while *The Candle Revisited* was the title of a collection of essays

[25] Michael Faraday, "A course of six lectures (adapted to a juvenile auditory), on the chemical history of a candle," *Chemical News* **3**: 6–10, 24–7, 42–6, 57–60, 72–6, 84–8, 1861.

[26] Crookes to Williams, in Frank A. J. L. James, "The letters of William Crookes to Charles Hanson Greville Williams 1861–2: the detection and isolation of thallium," *Ambix* **28**: 131–157, letter 17, 1981. See also "Michael Faraday," *Illustrated London News* **38**: 28–9 on p. 29, 1861.

[27] Michael Faraday, *Histoire d'une chandelle*, Hetzel, Paris, 1865; *Dzieje świecy*, Prószyński i S-ka, Warsaw, 1997; *Rousoku no Kagaku*, Oubunsha, Tokyo, 1969; *The Chemical History of a Candle. Put into Basic English by Phyllis Rossiter*, K. Paul, London, 1933.

[28] Jane Gregory and Steve Miller, *Science in Public: Communication, Culture, and Credibility*, Plenum Press, New York and London, 1998, pp. 133–6.

edited by a recent Director of The Royal Institution, Peter Day, and his colleague Richard Catlow.[29]

Of the 178 series delivered so far (no lectures were held in the 1939–1940 season and they did not resume until 1943–1944), 50 or just over a quarter have been published in book form since the 1860s. The subjects of the published lectures are not a simple reflection of the topics covered by the lectures. For example, nearly a fifth of all the series have been on chemistry, but only three books have been based on those lectures. However, roughly a quarter of the lectures have been devoted to natural philosophy or physics, which is in line with the proportion of published texts derived from those lectures; this preponderance is reflected in this volume. Some subject areas have been little covered, either in the lectures or in books. For example, the earth sciences have only had eight series, while mathematics has had three. Christopher Zeeman's 1978–1979 series on mathematics was the first time the subject was covered, but it was particularly significant in that it led to the establishment of the highly successful Royal Institution Mathematics Masterclasses throughout the country. In addition to the four lectures on natural philosophy or physics included in this volume, each of the major branches of science is represented by one lecture. The only major gap is that it has not been possible to include a lecture on engineering or technology, which accounts for just over 10% of the Christmas Lecture series, although the lecture by Lawrence Bragg does include a significant discussion on electrical engineering. This anthology thus provides a very small, but reasonably representative, sample of published Christmas Lectures from the 1860s until the 1990s. The lectures published here illustrate not only some commonalities over time and between lecturers, but also the differences in approaches to writing popular science books that can be taken in terms of level, language, tone, style, and so on.

---

[29] Peter Day and C. R. A. Catlow (eds.), *The Candle Revisited: Essays on Science and Technology*, Oxford University Press, Oxford, 1994.

Some lecturers, such as both Braggs in this volume, followed the pattern of publication established by Faraday; but others such as Tyndall, Robert Stawell Ball (1840–1913), and Ian Stewart divided or combined material from their lectures (and elsewhere) in different ways; while others declined to publish at all, though whether out of principle or lack of time is impossible to say. The case of Tyndall (who greatly admired Faraday) is particularly interesting, since he evidently thought that producing six longish chapters based explicitly on six lectures was perhaps not the best way of communicating science; instead, he divided the publication of his Christmas Lectures into small chunks of text, two examples of which are published here in the unproven hope that they might bear some relationship to the lectures as delivered. On the other hand, Ball combined two series of lectures (1881–1882, 1887–1888) into a single book, while Stewart added material to his book that was not delivered in the theatre (the material in the chapter published here was delivered as a Christmas Lecture).

The televising of the Christmas Lectures arose from the desire of The Royal Institution to make its work better known. In 1959, Lawrence Bragg made a series of six 15-minute television programmes entitled "The Nature of Things". This was more or less the title of his father's 1923–1924 Christmas Lectures, of which the first lecture is published here. Although Lawrence Bragg's lectures attracted an audience of about four million (it is not clear whether this figure was the aggregate or per lecture) and were declared a great success,[30] it was not until Laithwaite's 1966–1967 series entitled "The Engineer in Wonderland" that annual televising of the Christmas Lectures commenced. This happened following the founding of the BBC2 television channel in 1964 and when its controller was David Attenborough, who later gave the 1973–1974 Christmas Lectures. With their annual televising over the Christmas

---

[30] *Record of the Royal Institution 1960*, pp. 129–30.

and New Year holiday, the Christmas Lectures began to reach audience sizes which dwarfed what was possible with the printed page.

The early series of televised lectures, whilst not precisely live, underwent very little editing; indeed, on one occasion, the tape was delivered for broadcasting only 1 hour before transmission.[31] Thus, the broadcast lectures had that frisson of excitement that something unscripted might happen (an experiment might fail, for example), although Coates's professional pride would not permit such a thing to happen, even if he had to "fix it". For two decades, Coates would appear on the nation's screen each Christmas, keeping the lectures going and making witty interjections. Arguably, his presence contributed significantly to the success of the televised lectures. The production values demanded by television had a number of effects on the lectures. The number in each series was reduced from six to five in the late 1980s; and as the time needed to prepare for the lectures increased, it became rare following Laithwaite for a scientist to deliver more than one series. Indeed, since 1966, only Laithwaite (1966–1967, 1974–1975), Porter (1969–1970, 1976–1977) and Taylor (1971–1972, 1989–1990) have given more than one series.

The variety of approaches pursued by lecturers in turning the spoken word, experimental demonstrations, and visual illustrations into printed form support Faraday's initial reservations about the problems of doing just that. To be sure, the fact that publishers have continued to publish the lectures, many of which have gone into later editions, shows that there is a strong desire to retain and disseminate them in some permanent form. Nevertheless, it is their televising, their connection with Faraday (whose name is invariably worked in by the lecturer each year), that clearly makes the lectures a national — indeed, global — institution continuing to make a significant contribution to communicating science to a large growing audience.

---

[31] Personal communication from Bill Coates.

Biographical Notes on Lecturers

Ball, Robert Stawell (1840–1913)
Christmas Lectures: 1881–1882, 1887–1888, 1892–1893, 1898–1899, 1900–1901

Born in Dublin, Ball was educated there and near Chester before studying at Trinity College, Dublin. After spending 2 years tutoring the children of the astronomer the Third Earl of Rosse, he was appointed in 1867 as a professor at the Royal College of Science in Dublin, a position which he held until he moved in 1874 to be Professor of Astronomy at Trinity College, Astronomer Royal for Ireland, and Director of the Dunsink Observatory. In 1892, he was appointed Professor of Astronomy at the University of Cambridge and Director of its observatory. As a professional astronomer, he was mainly concerned with issues relating to instrumentation and measuring stellar parallax. As a popular lecturer in astronomy, he was in constant demand in Britain, Ireland, and the United States. Indeed, he was able to make himself relatively wealthy from his lecture fees.

Bragg, William Henry (1862–1942)
Christmas Lectures: 1919–1920, 1923–1924, 1925–1926, 1931–1932

Born near Wigton, Bragg attended King William's College in the Isle of Man before studying mathematics at Trinity College, Cambridge.

He was appointed Professor of Mathematics and Physics at the University of Adelaide in 1886. In the mid-1890s, he made the first X-ray tube in Australia, and started work on the theory and use of X-rays. In 1909, he returned to England to become Professor of Physics at the University of Leeds; and there, with his son William Lawrence Bragg, worked out how to determine the molecular structure of crystals using X-rays, for which they were jointly awarded the Nobel Prize for Physics in 1915. That year, Bragg was appointed Quain Professor of Physics at University College, London, although he spent much of the Great War working for the Admiralty on the acoustic detection of submarines. Unhappy at University College, he moved to The Royal Institution in 1923, where he remained until he died. In 1928 he was President of the British Association, and President of the Royal Society from 1935 to 1940. In that latter capacity in the late 1930s, he played a major role in preparing for the mobilisation of scientists for the 1939–1945 war and was also active in helping academics flee Fascist regimes to find new positions.

Bragg, William Lawrence (1890–1971)
Christmas Lectures: 1934–1935, 1961–1962

The son of William Henry Bragg, William Lawrence Bragg was born in Adelaide, where he attended St Peter's College and studied science at the University of Adelaide. When the family moved to England in 1909, he studied mathematics and physics at Trinity College, Cambridge. He and his father, in Leeds, worked out how to determine the molecular structure of crystals using X-rays, for which they jointly received the Nobel Prize in 1915; at the age of 25, Lawrence Bragg remains the youngest ever winner of the Prize. He served as an officer, rising to the rank of Major, in the Army in France throughout the Great War, where he worked on using sound to determine the location of enemy guns. In 1919, he succeeded Ernest Rutherford as Professor of Physics at the University of Manchester, where he continued his work on crystallography.

Between 1937 and 1938, he was Director of the National Physical Laboratory before again succeeding Rutherford as Professor of Physics at the University of Cambridge and Director of the Cavendish Laboratory. During the 1939–1945 war, he served on a number of government scientific committees. At the Cavendish, he established the Unit for the Study of Molecular Structure of Biological Systems funded by the Medical Research Council where, under his direction, Max Perutz, John Kendrew, Francis Crick, and James Watson carried out much of their seminal work on the structures of proteins and DNA. Following a bitter row at The Royal Institution, he moved there at the start of 1954 to address the issues that had created the problems. He was successful in resolving many of them and in 1965 was appointed the first Director of The Royal Institution, retiring the following year.

Faraday, Michael (1791–1867)

Christmas Lectures: 1827–1828, 1829–1830, 1832–1833, 1835–1836, 1837–1838, 1841–1842, 1843–1844, 1845–1846, 1848–1849, 1851–1852, 1852–1853, 1853–1854, 1854–1855, 1855–1856, 1856–1857, 1857–1858, 1858–1859, 1859–1860, 1860–1861

Faraday was born in Newington Butts, Southwark, the son of a blacksmith who came from the northwest of England. His father was a member of the small Sandemanian sect of Christianity; and Faraday became fully committed to this, making his confession of faith in 1821, serving as a deacon in the 1830s and as an elder in the 1840s and again in the 1860s. He served an apprenticeship with George Riebau as a bookbinder from 1805 to 1812. He was an assistant in The Royal Institution's laboratory for part of 1813 and again from 1815 to 1826 (touring the continent with Humphry Davy in the interim). He was appointed Assistant Superintendent of the House of The Royal Institution in 1821, Director of the Laboratory in 1825, and 8 years later the Fullerian Professorship of Chemistry was created for him. He was appointed Scientific Adviser to the

Admiralty in 1829; Professor of Chemistry at the Royal Military Academy, Woolwich, between 1830 and 1852; and Scientific Adviser to Trinity House from 1836 to 1865. His major discoveries include electromagnetic rotations (1821); benzene (1825); electromagnetic induction (1831); the laws of electrolysis and coining words such as electrode, cathode, and ion (early 1830s); the magneto-optical effect and diamagnetism (both 1845); and thereafter the establishment of the field theory of electromagnetism. He was twice offered the Presidency of the Royal Society, but declined on both occasions. He publicly stated several times that he would not accept a knighthood, but no evidence has been found that he was ever offered one. He was, however, awarded a Civil List Pension in 1836, and in 1858 Queen Victoria provided him with a Grace and Favour House at Hampton Court where he died.

Gregory, Richard Langton (b. 1923)
Christmas Lectures: 1967–1968

Born in London, Gregory attended King Alfred School, Hampstead, before serving in the signals section of the Royal Air Force (RAF) during much of the 1939–1945 war. At the instigation of the Air Ministry, he spent some of 1946 explaining radar and electronic communications to the general public at the Oxford Street bomb site where John Lewis is now located. From 1947 to 1950, he read philosophy and experimental psychology at Downing College, Cambridge. He spent the next 17 years in Cambridge holding a number of positions, researching perceptual issues, and publishing in 1966 his influential book *Eye and Brain*. In 1967, he cofounded the Department of Machine Intelligence and Perception at the University of Edinburgh. Three years later, he moved to Bristol as Professor of Neuropsychology as well as Director of the Brain and Perception Laboratory, where he remained until retirement. Always committed to promoting science to a broad audience, he was one of the regular judges on the BBC TV series "Young Scientist of the Year" in the 1970s. But, perhaps most significantly in terms of promoting science, Gregory, as a direct result of the findings of his research, founded in Bristol in 1981 the Exploratory hands-on science centre, the first such centre in Britain. From its opening in 1987 until its closure in 1999, it was visited by more than two million people. They saw and handled some highly innovative exhibits helping them understand basic scientific principles.

Lankester, Edwin Ray (1847–1929)
Christmas Lectures: 1903–1904

Born in London, Lankester attended St Paul's School before studying for a year at Downing College, Cambridge. He then moved to Christ Church, Oxford, where he studied natural science. After graduating, he travelled around the continent and worked in 1871 and 1872 at the Naples Biological Laboratory. He returned to Oxford as a tutor at Exeter College, but in 1875 moved to London as Professor of Zoology at University College, a position which he held until 1891 when he became Professor of Comparative Anatomy at the University of Oxford. He reorganised the museum there and with that experience was appointed Director of the British Museum (Natural History) in South Kensington in 1898, which he held until retirement in 1907. Between 1898 and 1901, he was also Fullerian Professor of Physiology and Comparative Anatomy at The Royal Institution. He was President of the British Association in 1906. Although he did not specialise in any one area of zoology, his work on embryology, parasitology, and fossil fishes exerted great influence.

Stewart, Ian (b. 1945)
Christmas Lectures: 1997–1998

Born in Folkestone, Stewart studied mathematics at Churchill College, Cambridge, and then at the University of Warwick where he received his PhD in 1969. He was then immediately appointed as a lecturer in mathematics. Promoted to Reader in 1984 and Professor in 1990, he has also held a number of positions overseas, and in 1994 was Gresham Professor of Geometry. An active research mathematician with over 170 published papers, much of his work is on chaos theory; he is particularly interested in problems that lie in the gaps between pure and applied mathematics. As a populariser of mathematics and related areas, he has written widely, publishing books such as _Does God Play Dice?_ (1989), _Letters to a Young Mathematician_ (2006), and _Why Beauty is Truth_ (2007). Stewart has written many science fiction stories, including, with Terry Pratchett and Jack Cohen, three books on _The Science of Discworld_.

Thompson, Silvanus Phillips (1851–1916)
Christmas Lectures: 1896–1997, 1910–1911

Born into a York Quaker family, Thompson attended Bootham School there. He took an external BA from the University of London, and in 1870 was appointed as a junior master at his old school. In 1875, he moved to London where he studied science at South Kensington, taking a University of London BSc degree. The following year, he was appointed as a physics lecturer at University College, Bristol, being promoted to Professor in 1878 after the award of a DSc from London. His interests centred on electricity and its practical applications, and in 1885 he returned to London as Principal of Finsbury Technical College where he remained for the rest of his life. The college was where a large number of late 19th and early 20th century electrical engineers learnt their profession. Thompson thus became a key figure in developing, in many ways, electrical engineering. As well as his scientific and engineering work, Thompson was interested in the history of science, writing useful biographies of Michael Faraday and Lord Kelvin. As a lifelong and very active Quaker, Thompson was outspoken in his criticisms of both the Boer War and the Great

War. His views probably damaged his career and his inability to save his young students and assistants at Finsbury Technical College from conscription into the trenches may have hastened his death.

Tyndall, John (1820–1893)

Christmas Lectures: 1861–1862, 1863–1864, 1865–1866, 1867–1868, 1869–1870, 1871–1872, 1873–1874, 1875–1876, 1877–1878, 1879–1880, 1882–1883, 1884–1885

Born in County Carlow, Tyndall attended the local National School there, and in 1839 joined the Ordnance Survey of Ireland. Three years later, he moved to the English Ordnance Survey where he worked until 1844, when he became a railway surveyor. In 1847, he was appointed to teach mathematics and surveying at Queenwood College, Hampshire. The following year, he moved to Marburg, where he spent 2 years studying with Robert Bunsen, taking his PhD in 1850; and the following year spent some time in Berlin. Thereafter, he returned to Queenwood College until his appointment at The Royal Institution as Professor of Natural Philosophy in 1853. In 1865, he succeeded Michael Faraday (whose first biographer he became) as Scientific Adviser to Trinity House, a position that he held until 1883 and in which capacity he worked on the transmission of sound. Following Faraday's death in 1867, Tyndall was appointed to his positions of Superintendent of the House and Director of the Laboratory. He retired from all of his positions in 1887. Tyndall undertook important work on diamagnetism, radiant heat, and spontaneous generation. He also worked on glaciers and became a keen mountaineer, being the first person, in 1861, to climb the Weisshorn. He wrote many popular scientific articles and books and engaged in polemics, most notably his Presidential Address to the 1874 meeting of the British Association at Belfast, espousing scientific naturalism.

The Correlation of the Physical Forces

Michael Faraday*

WE have frequently seen, during the course of these lectures, that one of those powers or forces of matter, of which I have written the names on that board, has produced results which are due to the action of some other force. Thus, you have seen the force of electricity acting in other ways than in attracting; you have also seen it combine matters together or disunite them by means of its action on the chemical force; and in this case, therefore, you have an instance in which these two powers are related. But we have other and deeper relations than these; we have not merely to see how it is that one power affects another—how the force of heat affects chemical affinity, and so forth, but we must try and comprehend what relation they bear to each other, and how these powers may be changed one into the other; and it will to-day require all my care, and your care too, to make this clear to your minds. I shall be obliged to confine myself to one or two instances, because to take in the whole extent of this mutual relation and conversion of forces would surpass the human intellect.

In the first place, then, here is a piece of fine zinc-foil, and if I cut it into narrow strips and apply to it the power of heat, admitting

* The original version appeared in: Michael Faraday, Lecture 6, *A Course of Six Lectures on the Various Forces of Matter and Their Relations to Each Other*, London, 1860, pp. 130–154.

the contact of air at the same time, you will find that it burns; and then, seeing that it burns, you will be prepared to say that there is chemical action taking place. You see all I have to do is to hold the piece of zinc at the side of the flame, so as to let it get heated, and yet to allow the air which is flowing into the flame from all sides to have access to it;—there is the piece of zinc burning just like a piece of wood, only brighter. A part of the zinc is going up into the air, in the form of that white smoke, and part is falling down on to the table. This, then, is the action of chemical affinity exerted between the zinc and the oxygen of the air. I will show you what a curious kind of affinity this is by an experiment, which is rather striking when seen for the first time. I have here some iron filings and gunpowder, and will mix them carefully together, with as little rough handling as possible; now we will compare the combustibility, so to speak, of the two. I will pour some spirit of wine into a basin and set it on fire: and, having our flame, I will drop this mixture of iron filings and gunpowder through it, so that both sets of particles will have an equal chance of burning. And now tell me which of them it is that burns?—you see a plentiful combustion of the iron filings; but I want you to observe, that though they have equal chances of burning, we shall find that by far the greater part of the gunpowder remains untouched; I have only to drain off this spirit of wine, and let the powder which has gone through the flame dry, which it will do in a few minutes, and I will then test it with a lighted match. So ready is the iron to burn, that it takes, under certain circumstances, even less time to catch fire than gunpowder. [As soon as the gunpowder was dry, Mr. Anderson handed it to the Lecturer, who applied a lighted match to it, when a sudden flash showed how large a proportion of gunpowder had escaped combustion when falling through the flame of alcohol.]

These are all cases of chemical affinity, and I show them to make you understand that we are about to enter upon the consideration of a strange kind of chemical affinity, and then to see how far we are enabled to convert this force of affinity into electricity or magnetism, or any other of the forces which we have discussed. Here is some zinc (I keep to the metal zinc as it is very useful for our purpose), and I can produce hydrogen gas by putting the zinc and sulphuric acid together, as they are in that retort; there you see the mixture which gives us hydrogen—the zinc is pulling the water to pieces and setting free hydrogen gas. Now we have learned by experience that if a little mercury is spread over that zinc, it does not *take away* its power of decomposing the water, but *modifies* it most curiously. See how that mixture is now boiling, but when I add a little mercury to it the gas ceases to come off. We have now scarcely a bubble of hydrogen set free, so that the action is suspended for the time. We have not *destroyed* the power of chemical affinity, but modified it in a wonderful and beautiful manner. Here are some pieces of zinc covered with mercury, exactly in the same way as the zinc in that retort is covered; and if I put this plate into sulphuric acid I get no gas, but this most extraordinary thing occurs, that if I introduce along with the zinc another metal which is *not* so combustible, then I reproduce all the action. I am now going to put to the amalgamated zinc in this retort some portions of copper wire (copper not being so combustible a metal as the zinc), and observe how I get hydrogen again, as in the first instance—there, the bubbles are coming over through the pneumatic trough, and ascending faster and faster in the jar; the zinc now is acting by reason of its contact with the copper.

Every step we are now taking brings us to a knowledge of new phenomena. That hydrogen which you now see coming off so abundantly does not come from the zinc as it did before, *but from the copper*. Here is a jar containing a solution of copper. If I put a piece of this amalgamated zinc into it, and leave it there, it has scarcely any action, and here is a plate of platinum which I will immerse in the same solution, and might leave it there for hours, days, months, or even years, and no action would take place. But by putting them both together and allowing them to touch (*fig.* 44), you see what a

Fig. 44.

Fig. 45.

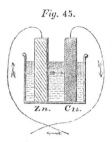

Zn. Cu.

coating of copper there is immediately thrown down on the platinum. Why is this? The platinum has no power of itself to reduce that metal from that fluid, but it has in some mysterious way received this power by its contact with the metal zinc. Here then you see a strange transfer of chemical force from one metal to another—the chemical force from the zinc is transferred, and made over to the platinum by the mere association of the two metals. I might take instead of the platinum, a piece of copper or of silver, and it would have no action of its own on this solution, but the moment the zinc was introduced and touched the other metal, then the action would take place, and it would become covered with copper. Now, is not this most wonderful and beautiful to see? We still have the identical chemical force of the particles of zinc acting, and yet in some strange manner we have power to make that chemical force, or something it produces, travel from one place to another—for we do make the chemical force travel from the zinc to the platinum by this very curious experiment of using the two metals in the same fluid in contact with each other.

Let us now examine these phenomena a little more closely. Here is a drawing (*fig.* 45) in which I have represented a vessel containing the acid liquid and the slips of zinc and platinum or copper, and I have shown them touching each other *outside* by means of a wire coming from each of them (for it matters not whether they touch in the fluid or outside—by pieces of metal attached, they still by that communication between them have this power transferred from one to the other). Now, if instead of only using one vessel, as I have shown there, I take another, and another, and put in zinc and platinum, zinc and platinum, zinc and platinum, and connect the platinum of one vessel with the zinc of another, the platinum of this vessel with the zinc of that, and so on, we should only be using a series of these vessels instead of one. This we have done in that arrangement which you see behind me. I am using what we call a Grove's voltaic battery, in which one metal is zinc, and the other platinum, and I have as many as forty pairs of these plates all exercising their force at once in sending the whole amount of chemical power there evolved through these wires under the floor and up to these two rods coming through the table. We need do no more than just bring these two ends in contact, when the spark shows us what power is present; and what a strange thing it is to see that this force is brought away from the battery behind me, and carried along through these wires. I have here an apparatus (*fig.* 46) which Sir Humphry Davy constructed many years ago, in order to see whether this power from the voltaic battery caused bodies to attract each other in the same manner as the ordinary electricity did. He made it in order to experiment with his large voltaic battery, which was the most powerful then in existence. You see there are in this glass jar two leaves of gold, which I can cause to move to and fro by this rack work. I will connect each of these gold leaves with separate ends of this battery, and if I have a sufficient number of plates in the battery I shall be able to show you that there will be some attraction between those leaves even before they come in contact: if I bring them sufficiently near when they are in communication with the ends of the battery, they will be drawn gently together, and you will know when this takes place, because the power will cause the gold leaves to burn away, which they could only do when they touched each other. Now I am going to cause these two leaves of gold to approach gradually, and I have no doubt that some of you will see that they approach before they burn, and those who are too far off to see them approach will see by their burning that they have come together. Now they are attracting each other, long before the connection is complete, and there they go! burnt up in that brilliant flash, so strong is the force. You thus see, from the attractive force at the two ends of this battery, that these are really and truly electrical phenomena.

Now, let us consider what is this spark. I take these two ends and bring them together, and there I get this glorious spark like the sunlight in the heavens above us. What is this? It is the same thing which you saw when I discharged the large electrical machine, when you saw one single bright flash; it is the same thing, only *continued*, because here we have a more effective arrangement. Instead of having a machine which we are obliged to turn for a long time together, we have here a *chemical* power which sends forth the spark — and it is wonderful and beautiful to see how this spark is carried about through these wires. I want you to perceive, if possible, that this very spark and the heat it produces (for there is heat), is neither more nor less than the chemical force of the zinc — its *very* force carried along wires and conveyed to this place. I am about to take a portion of the zinc and burn it in oxygen gas for the sake of showing you the kind of light produced by the actual combustion in oxygen gas of some of this metal. [A tassel of zinc-foil was ignited at a spirit-lamp and introduced into a jar of oxygen, when it burnt with a brilliant light.] That shows you what the affinity is when we come to consider it in its energy and power. And the zinc is being burned in the battery behind me at a much more rapid rate than you see in that jar, because the zinc is there dissolving and *burning*, and produces here this great electric light. That very same power which in that jar you saw evolved from the actual combustion of the zinc in oxygen, is carried along these wires and made evident here, and you may if you please consider that the zinc is burning in those cells, and that *this* is the light of that burning [bringing the two poles in contact and showing the electric light]; and we might so arrange our apparatus as to show that the amounts of power evolved in either case are identical. Having thus obtained power over the chemical force, how wonderfully we are able to convey it from place to place! When we use gunpowder for

Fig. 46.

Fig. 47.

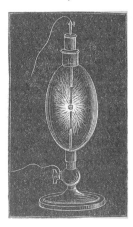

explosive purposes, we can send into the mine chemical affinity by means of this electricity; not having provided fire beforehand, we can send it in at the moment we require it. Now here (*fig.* 47) is a vessel containing two charcoal points, and I bring it forward as an illustration of the wonderful power of conveying this force from place to place. I have merely to connect these by means of wires to the opposite ends of the battery, and bring the points in contact. See what an exhibition of force we have! We have exhausted the air so that the charcoal cannot burn, and therefore the light you see is really the burning of the zinc in the cells behind me — there is no disappearance of the carbon, although we have that glorious electric light; and the moment I cut off the connection it stops. Here is a better instance to enable some of you to see the certainty with which we can convey this force, where, under ordinary circumstances, chemical affinity would not act. We may absolutely take these two

charcoal poles down under water, and get our electric light there; —
there they are in the water, and you observe when I bring them into
connection we have the same light as we had in that glass vessel.

Now, besides this production of light we have all the other
effects and powers of burning zinc. I have a few wires here which
are not combustible, and I am going to take one of them, a small
platinum wire, and suspend it between these two rods which are

Fig. 48.

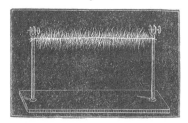

connected with the battery, and when contact is made at the battery see what heat we get (*fig*. 48). Is not that beautiful?—it is a
complete bridge of power. There is metallic connection all the way
round in this arrangement, and where I have inserted the platinum, which offers some resistance to the passage of the force, you
see what an amount of heat is evolved,—this is the heat which the
zinc would give if burnt in oxygen, but as it is being burnt in the
voltaic battery it is giving it out at this spot. I will now shorten this
wire for the sake of showing you that the shorter the obstructing
wire is, the more and more intense is the heat, until at last our platinum is fused and falls down, breaking off the circuit.

Here is another instance. I will take a piece of the metal silver,
and place it on charcoal connected with one end of the battery, and
lower the other charcoal pole on to it. See how brilliantly it burns
(*fig*. 49). Here is a piece of iron on the charcoal, see what a combustion is going on; and we might go on in this way burning
almost everything we place between the poles. Now I want to

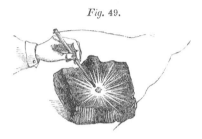

Fig. 49.

show you that this power is still chemical affinity—that if we call the power which is evolved at this point *heat*, or *electricity*, or any other name referring to its source, or the way in which it travels, we still shall find it to be chemical action. Here is a coloured liquid which can show by its change of colour the effects of chemical action; I will pour part of it into this glass and you will find that these wires have a very strong action. I am not going to show you any effects of combustion or heat, but I will take these two platinum plates, and fasten one to the one pole and the other to the other end, and place them in this solution, and in a very short time you will see the blue colour will be entirely destroyed. See, it is colourless now!—I have merely brought the end of the wires into the solution of indigo, and the power of electricity has come through these wires and made itself evident by its chemical action. There is also another curious thing to be noticed now we are dealing with the chemistry of electricity, which is that the chemical power which destroys the colour is only due to the action on one side. I will pour some more of this sulphindigotic acid[23] into a flat dish and will then make a porous dyke of sand separating the two

---

[23] *Sulphindigotic acid*. A mixture of one part of indigo and fifteen parts of concentrated oil of vitriol. It is bleached on the side at which hydrogen gas is evolved in consequence of the liberated hydrogen withdrawing oxygen from the indigo, thereby forming a colourless deoxidised indigo. In making the experiment, only enough of the sulphindigotic acid must be added to give the water a decided blue colour.

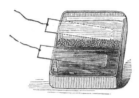

Fig. 50.

portions of fluid into two parts (*fig.* 50), and now we shall be able
to see whether there is any difference in the two ends of the battery, and which it is that possess this peculiar action. You see it is
the one on my right hand which has the power of destroying the
blue, for the portion on that side is thoroughly bleached, while
nothing has apparently occurred on the other side. I say *apparently*, for you must not imagine, that because you cannot perceive
any action none has taken place.

Here we have another instance of chemical action. I take these
platinum plates again and immerse them in this solution of copper
from which we formerly precipitated some of the metal, when the
platinum and zinc were both put in it together. You see that these
two platinum plates have no chemical action of any kind, they
might remain in the solution as long as I liked, without having any
power of themselves to reduce the copper; but the moment I bring
the two poles of the battery in contact with them, the chemical
action which is there transformed into electricity and carried along
the wires, again becomes chemical action at the two platinum
poles, and now we shall have the power appearing on the left hand
side, and throwing down the copper in the metallic state on the
platinum plate; and in this way I might give you many instances of
the extraordinary way in which this chemical action or electricity
may be carried about. That strange nugget of gold, of which there is
a model in the other room, and which has an interest of its own in
the natural history of gold, and which came from Ballarat, and was

worth 8000*l.* or 9000*l.* when it was melted down last November, was brought together in the bowels of the earth, perhaps ages and ages ago, by some such power as this. And there is also another beautiful result dependent upon chemical affinity in that fine lead-tree[24], the lead growing and growing by virtue of this power. The lead and the zinc are combined together in a little voltaic arrangement, in a manner far more important than the powerful one you see here, because in nature these minute actions are going on for ever, and are of great and wonderful importance in the precipitation of metals and formation of mineral veins, and so forth. These actions are not for a limited time, like my battery here, but they act for ever in small degrees, accumulating more and more of the results.

I have here given you all the illustrations that time will permit me to show you of chemical affinity producing electricity, and electricity again becoming chemical affinity. Let that suffice for the present; and let us now go a little deeper into the subject of this chemical force, or this electricity—which shall I name first—the one producing the other in a variety of ways. These forces are also wonderful in their power of producing another of the forces we have been considering, namely, that of magnetism, and you know that it is only of late years, and long since I was born, that the discovery of the relations of these two forces of electricity and chemical affinity to produce magnetism have become known. Philosophers had been suspecting this affinity for a long time, and had long had great hopes of success—for in the pursuit of science we first start with hopes and expectations; these we realise and

---

[24] *Lead tree.* To make a lead tree, pass a bundle of brass wires through the cork of a bottle, and fasten a plate of zinc round them just as they issue from the cork, so that the zinc may be in contact with every one of the wires. Make the wires to diverge so as to form a sort of cone, and having filled the bottle quite full of a solution of sugar of lead, insert the wires and cork and seal it down, so as to perfectly exclude the air. In a short time the metallic lead will begin to crystallise around the divergent wires, and form a beautiful object.

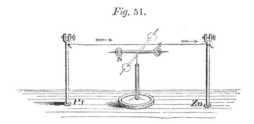

Fig. 51.

establish never again to be lost, and upon them we found new expectations of further discoveries, and so go on pursuing, realising, establishing, and founding new hopes again and again.

Now observe this: here is a piece of wire which I am about to make into a bridge of force, that is to say, a communicator between the two ends of the battery. It is copper wire only, and is therefore not magnetic of itself. We will examine this wire with our magnetic needle (*fig.* 51), and though connected with one extreme end of the battery, you see that before the circuit is completed it has no power over the magnet. But observe it when I make contact; watch the needle, see how it is swung round, and notice how indifferent it becomes if I break contact again; so you see we have this wire evidently affecting the magnetic needle under these circumstances. Let me show you that a little more strongly. I have here a quantity of wire which has been wound into a spiral, and this will affect the magnetic needle in a very curious manner, because, owing to its shape, it will act very like a real magnet. The copper spiral has no power over that magnetic needle at present; but if I cause the electric current to circulate through it, by bringing the two ends of the battery in contact with the ends of the wire which forms the spiral, what will happen? Why one end of the needle is most powerfully drawn to it; and if I take the other end of the needle it is repelled; so you see I have produced exactly the same phenomena as I had with the bar magnet,—one end attracting and the other repelling. Is not this then curious to see that we can construct a magnet of copper? Furthermore, if I take an iron bar, and put it inside the coil, so long as there is no electric current circulating round, it has no attraction,—as you will observe if I bring a little iron filings or nails near the iron. But now if I make contact with the battery they are attracted at once. It becomes at once a powerful magnet, so much so that I should not wonder if these magnetic needles on different parts of the table pointed to it. And I will show you by another experiment what an attraction it has. This piece and that piece of

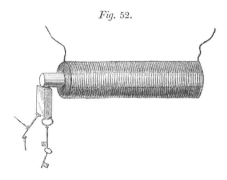

Fig. 52.

iron and many other pieces are now strongly attracted (*fig.* 52), but as soon as I break contact the power is all gone and they fall. What then can be a better or a stronger proof than this of the relation of the powers of magnetism and electricity? Again, here is a little piece of iron which is not yet magnetised. It will not at present take up any one of these nails; but I will take a piece of wire and coil it round the iron (the wire being covered with cotton in every part it does not touch the iron), so that the current must go round in this spiral coil—I am, in fact, preparing an *electro-magnet* (we are obliged to use such terms to express our meaning, because it is a magnet made by electricity,—because we produce by the force of electricity a magnet of far greater power than a permanent steel one). It is now completed and I will repeat the experiment which you saw the other day, of building up a bridge of iron nails; the

Fig. 53.

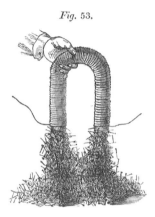

contact is now made and the current is going through; it is now a powerful magnet; here are the iron nails which we had the other day, and now I have brought this magnet near them they are clinging so hard that I can scarcely move them with my hand (*fig.* 53). But when the contact is broken, see how they fall. What can show you better than such an experiment as this the magnetic attraction with which we have endowed these portions of iron? Here again is a fine illustration of this strong power of magnetism. It is a magnet of the same sort as the one you have just seen. I am about to make the current of electricity pass through the wires which are round this iron for the purpose of showing you what powerful effects we get. Here are the poles of the magnet; and let us place on one of them this long bar of iron. You see as soon as contact is made how it rises in position (*fig.* 54); and if I take such a piece as this cylinder, and place it on, woe be to me if I get my finger between; I can roll it over, but if I try to pull it off, I might lift up the whole magnet, but I have no power to overcome the magnetic power which is here evident. I might give you an infinity of illustrations of this high magnetic power. There is that long bar of iron held out, and I have no doubt that if I were to examine the other end I should find that it was a magnet. See what power it must have to support not only

Fig. 54.

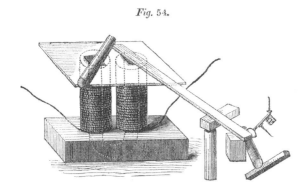

these nails, but all those lumps of iron, hanging on to the end. What then can surpass these evidences of the change of chemical force into electricity, and electricity into magnetism? I might show you many other experiments whereby I could obtain electricity and chemical action, heat and light from a magnet, but what more need I show you to prove the universal correlation of the physical forces of matter, and their mutual conversion one into another?

And now let us give place as juveniles to the respect we owe to our elders; and for a time let me address myself to those of our seniors who have honoured me with their presence during these lectures. I wish to claim this moment for the purpose of tendering our thanks to them, and my thanks to you all for the way in which you have borne the inconvenience that I at first subjected you to. I hope that the insight which you have here gained into some of the laws by which the universe is governed, may be the occasion of some amongst you turning your attention to these subjects; for what study is there more fitted to the mind of man than that of the physical sciences? And what is there more capable of giving him an insight into the actions of those laws, a knowledge of which gives interest to the most trifling phenomenon of nature, and makes the observing student find

"——tongues in trees, books in the running brooks,
Sermons in stones, and good in everything"?

Carbon or Charcoal—Coal Gas—Respiration and Its Analogy to the Burning of a Candle—Conclusion

Michael Faraday*

A LADY who honours me by her presence at these lectures, has conferred a still further obligation by sending me these two candles, which are from Japan, and, I presume, are made of that substance to which I referred in a former lecture. You see that they are even far more highly ornamented than the French candles, and, I suppose, are candles of luxury, judging from their appearance. They have a remarkable peculiarity about them; namely, a hollow wick,—that beautiful peculiarity which Argand introduced into the lamp and made so valuable. To those who receive such presents from the East, I may just say that this and such like materials, gradually undergo a change which gives them on the surface a dull and dead appearance; but they may easily be restored to their original beauty if the surface be rubbed with a clean cloth or silk handkerchief, so as to polish the little rugosity or roughness: this will restore the beauty of the colours. I have so rubbed one of these candles, and you see the difference between it and the other which has not been polished, but which may be restored by the same process. Observe, also, that these moulded candles from Japan are made more conical than the moulded candles in this part of the world.

---

* The original version appeared in: Michael Faraday, Lecture 6, *A Course of Six Lectures on the Chemical History of a Candle*, London, 1861, pp. 142–171.

I told you, when we last met, a good deal about carbonic acid. We found by the lime-water test, that when the vapour from the top of the candle or lamp was received into bottles and tested by this solution of lime-water (the composition of which I explained to you, and which you can make for yourselves), we had that white opacity which was in fact calcareous matter, like shells and corals, and many of the rocks and minerals in the earth. But I have not yet told you fully and clearly the chemical history of this substance, carbonic acid, as we have it from the candle, and I must now resume that subject. We have seen the products, and the nature of them, as they issue from the candle. We have traced the water to its elements, and now we have to see where are the elements of the carbonic acid supplied by the candle: a few experiments will show this. You remember that when a candle burns badly it produces smoke; but if it is burning well, there is no smoke. And you know that the brightness of the candle is due to this smoke, which becomes ignited. Here is an experiment to prove this: so long as the smoke remains in the flame of the candle and becomes ignited it gives a beautiful light, and never appears to us in the form of black particles. I will light some fuel, which is extravagant in its burning; this will serve our purpose—a little turpentine on a sponge. You see the smoke rising from it, and floating into the air in large quantities; and remember now, the carbonic acid that we have from the candle is from such smoke as that. To make that evident to you, I will introduce this turpentine burning on the sponge into a flask where I have plenty of oxygen, the rich part of the atmosphere, and you now see that the smoke is all consumed. This is the first part of our experiment, and now what follows? The carbon which you saw flying off from the turpentine flame in the air is now entirely burned in this oxygen, and we shall find that it will, by this rough and temporary experiment, give us exactly the same conclusion and result as we had from the combustion of the candle. The reason why I make the experiment in this manner is solely that I may cause the steps of our demonstration to be so simple that you can never for a moment lose the train of reasoning, if you only pay attention. All the carbon which is burned in oxygen, or air, comes out as carbonic acid, whilst those particles which are not so burned show you the second substance in the carbonic acid; namely, the carbon—that body which made the flame so bright whilst there was plenty of air, but which was thrown off in excess when there was not oxygen enough to burn it.

I have also to show you a little more distinctly, the history of carbon and oxygen in their union to make carbonic acid. You are now better able to understand this than before, and I have prepared three or four experiments by way of illustration. This jar is filled with oxygen, and here is some carbon which has been placed in a crucible, for the purpose of being made red-hot. I keep my jar dry, and venture to give you a result imperfect in some degree, in order that I may make the experiment brighter. I am about to put the oxygen and the carbon together. That this is carbon (common charcoal pulverized) you will see by the way in which it burns in the air [letting some of the red-hot charcoal fall out of the crucible]. I am now about to burn it in oxygen gas, and look at the difference. It may appear to you at a distance as if it were burning with a flame; but it is not so. Every little piece of charcoal is burning as a spark, and whilst it so burns it is producing carbonic acid. I specially want these two or three experiments to point out what I shall dwell upon more distinctly by and by—that carbon burns in this way, and not as a flame.

Instead of taking many particles of carbon to burn I will take a rather large piece, which will enable you to see the form and size, and to trace the effects very decidedly. Here is the jar of oxygen, and here is the piece of charcoal, to which I have fastened a little piece of wood, which I can set fire to, and so commence the combustion, which I could not conveniently do without. You now see the charcoal burning, but not as a flame (or if there be a flame it is the smallest possible one, which I know the cause of; namely, the formation of a little carbonic oxide close upon the surface of the carbon). It goes on burning, you see, slowly producing carbonic acid by the union of this carbon or charcoal (they are equivalent terms) with the oxygen. I have here another piece of charcoal, a piece of bark, which has the quality of being blown to pieces—exploding—as it burns. By the effect of the heat we shall reduce the lump of carbon into particles that will fly off; still every particle, equally with the whole mass, burns in this peculiar way—it burns as a coal and not like a flame. You observe a multitude of little combustions going on, but no flame. I do not know a finer experiment than this to show that carbon burns with a spark.

Here, then, is carbonic acid formed from its elements. It is produced at once; and if we examined it by lime-water, you will see that we have the same substance which I have previously described to you. By putting together 6 parts of carbon by weight (whether it comes from the flame of a candle or from powdered charcoal) and 16 parts of oxygen by weight, we have 22 parts of carbonic acid; and, as we saw last time, the 22 parts of carbonic acid combined with 28 parts of lime, produced common carbonate of lime. If you were to examine an oyster-shell and weigh the component parts, you would find that every 50 parts would give 6 of carbon and 16 of oxygen combined with 28 of lime. However, I do not want to trouble you with these minutiæ; it is only the general philosophy of the matter that we can now go into. See how finely the carbon is dissolving away [pointing to the lump of charcoal burning quietly in the jar of oxygen]. You may say that the charcoal is actually dissolving in the air round about; and if that were perfectly pure charcoal, which we can easily prepare, there would be no residue whatever. When we have a perfectly cleansed and purified piece of carbon, there is no ash left. The carbon burns as a solid dense body, that heat alone cannot change as to its solidity, and yet it passes away into vapour that never condenses into solid or liquid under ordinary circumstances; and what is more curious still is the fact that the oxygen does not change in its bulk by the solution of the carbon in it. Just as the bulk is at first, so it is at last, only it has become carbonic acid.

There is another experiment which I must give you before you are fully acquainted with the general nature of carbonic acid. Being a compound body, consisting of carbon and oxygen, carbonic acid is a body that we ought to be able to take asunder. And so we can. As we did with water, so we can with carbonic acid,—take the two parts asunder. The simplest and quickest way is to act upon the carbonic acid by a substance that can attract the oxygen from it, and leave the carbon behind. You recollect that I took potassium and put it upon water or ice, and you saw that it could take the oxygen from the hydrogen. Now, suppose we do something of the same kind here with this carbonic acid. You know carbonic acid to be a heavy gas: I will not test it with lime-water, as that will interfere with our subsequent experiments, but I think the heaviness of the gas and the power of extinguishing flame will be sufficient for our purpose. I introduce a flame into the gas, and you will see whether it will be put out. You see the light is extinguished. Indeed, the gas may, perhaps, put out phosphorus, which you know has a pretty strong combustion. Here is a piece of phosphorus heated to a high degree. I introduce it into gas, and you observe the light is put out, but it will take fire again in the air, because there it re-enters into combustion. Now let me take a piece of potassium, a substance which even at common temperatures can act upon carbonic acid, though not sufficiently for our present purpose, because it soon gets covered with a protecting coat; but if we warm it up to the burning point in air, as we have a fair right to do, and as we have done with phosphorus, you will see that it can burn in carbonic acid; and if it burns, it will burn by taking oxygen, so that you will see what is left behind. I am going, then, to burn this potassium in the carbonic acid, as a proof of the existence of oxygen in the carbonic acid. [In the preliminary process of heating the potassium exploded.] Sometimes we get an awkward piece of potassium that explodes, or something like it, when it burns. I will take another piece, and now that it is heated I introduce it into the jar, and you perceive that it burns in the carbonic acid—not so well as in the air, because the carbonic acid contains the oxygen combined, but it does burn, and takes away the oxygen. If I now put this potassium into water, I find that besides the potash formed (which you need not trouble about) there is a quantity of carbon produced. I have here made the experiment in a very rough way, but I assure you that if I were to make it carefully, devoting a day to it, instead of five minutes, we should get all the proper amount of charcoal left in the spoon, or in the place where the potassium was burned, so that there could be no doubt as to the result. Here, then, is the carbon obtained from the carbonic acid, as a common black substance; so that you have the entire proof of the nature of carbonic acid as consisting of carbon and oxygen. And now, I may tell you, that *whenever* carbon burns under common circumstances, it produces carbonic acid.

Suppose I take this piece of wood, and put it into a bottle with lime-water. I might shake that lime-water up with wood and the atmosphere as long as I pleased, it would still remain clear as you see it; but suppose I burn the piece of wood in the air of that bottle. You, of course, know I get water. Do I get carbonic acid? [The experiment was performed.] There it is, you see—that is to say, the carbonate lime, which results from carbonic acid, and that carbonic acid must be formed from the carbon which comes from the wood, from the candle, or any other thing. Indeed, you have yourselves frequently tried a very pretty experiment, by which you may see the carbon in wood. If you take a piece of wood, and partly burn it, and then blow it out, you have carbon left. There are things that do not show carbon in this way. A candle does not so show it, but it contains carbon. Here also is a jar of coal-gas, which produces carbonic acid abundantly,—you do not see the carbon, but we can soon show it to you. I will light it, and as long as there is any gas in this cylinder it will go on burning. You see no carbon, but you see a flame, and because that is bright it will lead you to guess that there is carbon in the flame. But I will show it to you by another process. I have some of the same gas in another vessel, mixed with a body that will burn the hydrogen of the gas, but will not burn the carbon. I will light them with a burning taper, and you perceive the hydrogen is consumed, but not the carbon, which is left behind as a dense black smoke. I hope that by these three or four experiments you will learn to see when carbon is present, and understand what are the products of combustion, when gas or other bodies are thoroughly burned in the air.

Before we leave the subject of carbon, let us make a few experiments and remarks upon its wonderful condition, as respects ordinary combustion. I have shown you that the carbon in burning burns only as a solid body, and yet you perceive that, after it is burned, it ceases to be a solid. There are very few fuels that act like this. It is in fact only that great source of fuel, the carbonaceous series, the coals, charcoals, and woods, that can do it. I do not know that there is any other elementary substance besides carbon that burns with these conditions; and if it had not been so, what would happen to us? Suppose all fuel had been like iron which, when it burns, burns into a solid substance. We could not then have such a combustion as you have in this fire-place. Here also is another kind of fuel which burns very well—as well as, if not better, than carbon—so well, indeed, as to take fire of itself when it is in the air, as you see. [Breaking a tube full of lead pyrophorus.] This substance is lead, and you see how wonderfully combustible it is. It is very much divided, and is like a heap of coals in the fire-place: the air can get to its surface and inside, and so it burns. But why does it not burn in that way now when it is lying in a mass? [Emptying the contents of the tube in a heap on to a plate of iron.] Simply because the air cannot get to it. Though it can produce a great heat, the great heat which we want in our furnaces and under our boilers, still that which is produced cannot get away from the portion which remains unburned underneath, and that portion, therefore, is prevented from coming in contact with the atmosphere, and cannot be consumed. How different is that from carbon! Carbon burns just in the same way as this lead does, and so gives an intense fire in the furnace, or wherever you choose to burn it; but then the body produced by its combustion passes away, and the remaining carbon is left clear. I showed you how carbon went on dissolving in the oxygen, leaving no ash; whereas, here [pointing to the heap of pyrophorus] we have actually more ash than fuel, for it is heavier by the amount of the oxygen which has united with it. Thus you see the difference between carbon and lead or iron: if we chose iron, which gives so wonderful a result in our applications of this fuel, either as light or heat. If, when the carbon burnt, the product went off as a solid body, you would have had the room filled with an opaque substance, as in the case of the phosphorus; but when carbon burns, everything passes up into the atmosphere. It is in a fixed, almost unchangeable condition before the combustion; but afterwards it is in the form of gas, which it is very difficult (though we have succeeded) to produce in a solid or liquid state.

Now I must take you to a very interesting part of our subject— to the relation between the combustion of a candle and that living kind of combustion which goes on within us. In every one of us there is a living process of combustion going on very similar to that of a candle, and I must try to make that plain to you. For it is not merely true in a poetical sense—the relation of the life of man to a taper; and if you will follow, I think I can make this clear. In order to make the relation very plain, I have devised a little apparatus which we can soon build up before you. Here is a board and a groove cut in it, and I can close the groove at the top part by a little cover; I can then continue the groove as a channel by a glass

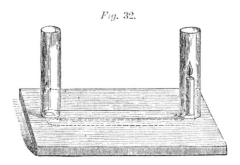

Fig. 32.

tube at each end, there being a free passage through the whole. Suppose I take a taper or candle (we can now be liberal in our use of the word "candle," since we understand what it means), and place it in one of the tubes; it will go on, you see, burning very well. You observe that the air which feeds the flame passes down the tube at one end, then goes along the horizontal tube, and ascends the tube at the other end in which the taper is placed. If I stop the aperture through which the air enters, I stop combustion, as you perceive. I stop the supply of air, and consequently the candle goes out. But now what will you think of this fact? In a former experiment I showed you the air going from one burning candle to a second candle. If I took the air proceeding from another candle, and sent it down by a complicated arrangement into this tube, I should put this burning candle out. But what will you say when I tell you that my breath will put out that candle? I do not mean by blowing at all, but simply that the nature of my breath is such that a candle cannot burn in it. I will now hold my mouth over the aperture, and without blowing the flame in any way, let no air enter the tube but what comes from my mouth. You see the result. I did not blow the candle out. I merely let the air which I expired pass into the aperture, and the result was that the light went out for want of oxygen, and for no other reason. Something or other—namely, my lungs—had taken away the oxygen from the air, and there was no more to supply the combustion of the candle. It is, I think, very pretty to see the time it takes before the bad air which I throw into this part of the apparatus has reached the candle. The candle at first goes on burning, but so soon as the air has had time to reach it it goes out. And now I will show you another experiment, because this is an important part of our philosophy. Here is a jar which contains fresh air, as you can see by the circumstance of a candle or gas-light burning it. I make it close for a little time, and by means of a pipe I get my mouth over it so that I can inhale the air. By putting it over water, in the way that you see, I am able to draw up this air (supposing the cork to be quite light), take it into my lungs, and throw it back into the jar: we can then examine it, and see the result. You observe, I first take up the air, and then throw it back, as is evident from the ascent and descent of the water, and now, by putting a taper into the air, you will see the state in which it is by the light being extinguished. Even one inspiration, you see, has completely spoiled this air, so that it is no use my trying to breathe it a second time. Now you understand the ground of the impropriety of many of the arrangements among the houses of the poorer classes, by which the air is breathed over and over again, for the want of a

Fig. 33.

Fig. 34.

supply, by means of proper ventilation, sufficient to produce a good result. You see how bad the air becomes by a single breathing; so that you can easily understand how essential fresh air is to us.

To pursue this a little further, let us see what will happen with lime-water. Here is a globe which contains a little lime-water, and it is so arranged as regards the pipes, as to give access to the air within, so that we can ascertain the effect of respired, or unrespired air upon it. Of course I can either draw in air (through A), and so make the air that feeds my lungs go through the lime-water, or I can force the air out of my lungs through the tube (B), which goes to the bottom, and so show its effect upon the lime-water. You will observe that however long I draw the external air into the lime-water, and then through it to my lungs, I shall produce no effect upon the water—it will not make the lime-water turbid; but if I throw the air *from* my lungs through the lime-water, several times in succession, you see how white and milky the water is getting, showing the effect which expired air has had upon it; and now you begin to know that the atmosphere which we have spoiled by respiration is spoiled by carbonic acid, for you see it here in contact with the lime-water.

I have here two bottles, one containing lime-water and the other common water, and tubes which pass into the bottles and connect them.

Fig. 35.

The apparatus is very rough, but it is useful notwithstanding. If I take these two bottles, inhaling here and exhaling there, the arrangement of the tubes will prevent the air going backwards. The air coming in will go to my mouth and lungs, and in going out, will pass through the lime-water, so that I can go on breathing and making an experiment, very refined in its nature, and very good in its results. You will observe that the good air has done nothing to the lime-water; in the other case nothing has come to the lime-water but my respiration, and you see the difference in the two cases.

Let us now go a little further. What is all this process going on within us which we cannot do without, either day or night, which is so provided for by the Author of all things that He has arranged that it shall be independent of all will? If we restrain our respiration, as we can to a certain extent, we should destroy ourselves. When we are asleep, the organs of respiration and the parts that are associated with them, still go on with their action, so necessary is this process of respiration to us, this contact of the air with the lungs. I must tell you, in the briefest possible manner, what this process is. We consume food: the food goes through that strange set of vessels and organs within us, and is brought into various

parts of the system, into the digestive parts especially; and alternately the portion which is so changed is carried through our lungs
by one set of vessels, while the air that we inhale and exhale is
drawn into and thrown out of the lungs by another set of vessels,
so that the air and the food come close together, separated only by
an exceedingly thin surface: the air can thus act upon the blood by
this process, producing precisely the same results in kind as we
have seen in the case of the candle. The candle combines with parts
of the air, forming carbonic acid, and evolves heat; so in the lungs
there is this curious, wonderful change taking place. The air entering, combines with the carbon (not carbon in a free state, but, as in
this case, placed ready for action at the moment), and makes carbonic acid, and is so thrown out into the atmosphere, and thus this
singular result takes place; we thus look upon the food as fuel. Let
me take that piece of sugar, which will serve my purpose. It is a
compound of carbon, hydrogen, and oxygen, similar to a candle, as
containing the same elements, though not in the same proportion;
the proportions being as shown in this table:—

SUGAR.

Carbon	72
Hydrogen	11
Oxygen	88

$\left.\begin{matrix} \\ \end{matrix}\right\} 99$

This is, indeed, a very curious thing, which you can well remember, for the oxygen and hydrogen are in exactly the proportions
which form water, so that sugar may be said to be compounded of
72 parts of carbon and 99 parts of water; and it is the carbon in the
sugar that combines with the oxygen carried in by the air in the
process of respiration, so making us like candles; producing these
actions, warmth, and far more wonderful results besides, for the
sustenance of the system, by a most beautiful and simple process.
To make this still more striking, I will take a little sugar; or to has-
ten the experiment I will use some syrup, which contains about

three-fourths of sugar and a little water. If I put a little oil of vitriol on it, it takes away the water, and leaves the carbon in a black mass. [The Lecturer mixed the two together.] You see how the carbon is coming out, and before long we shall have a solid mass of charcoal, all of which has come out of sugar. Sugar, as you know, is food, and here we have absolutely a solid lump of carbon where you would not have expected it. And if I make arrangements so as to oxidize the carbon of sugar, we shall have a much more striking result. Here is sugar, and I have here an oxidizer—a quicker one than the atmosphere; and so we shall oxidize this fuel by a process different from respiration in its form, though not different in its kind. It is the combustion of the carbon by the contact of oxygen which the body has supplied to it. If I set this into action at once, you will see combustion produced. Just what occurs in my lungs— taking in oxygen from another source, namely, the atmosphere, takes place here by a more rapid process.

You will be astonished when I tell you what this curious play of carbon amounts to. A candle will burn some four, five, six, or seven hours. What then must be the daily amount of carbon going up into the air in the way of carbonic acid! What a quantity of carbon must go from each of us in respiration! What a wonderful change of carbon must take place under these circumstances of combustion or respiration! A man in twenty-four hours converts as much as seven ounces of carbon into carbonic acid; a milch cow will convert seventy ounces, and a horse seventy-nine ounces, solely by the act of respiration. That is, the horse in twenty-four hours burns seventy-nine ounces of charcoal, or carbon, in his organs of respiration to supply his natural warmth in that time. All the warm-blooded animals get their warmth in this way, by the conversion of carbon, not in a free state, but in a state of combination. And what an extraordinary notion this gives us of the alterations going on in our atmosphere. As much as 5,000,000 pounds, or 548 tons, of carbonic acid is formed by respiration in London alone in twenty-four hours.

And where does all this go? Up into the air. If the carbon had been like the lead which I showed you, or the iron which, in burning, produces a solid substance, what would happen? Combustion could not go on. As charcoal burns it becomes a vapour and passes off into the atmosphere, which is the great vehicle, the great carrier for conveying it away to other places. Then what becomes of it? Wonderful is it to find that the change produced by respiration, which seems so injurious to us (for we cannot breathe air twice over), is the very life and support of plants and vegetables that grow upon the surface of the earth. It is the same also under the surface, in the great bodies of water; for fishes and other animals respire upon the same principle, though not exactly by contact with the open air.

Such fish as I have here [pointing to a globe of gold-fish] respire by the oxygen which is dissolved from the air by the water, and form carbonic acid, and they all move about to produce the one great work of making the animal and vegetable kingdoms sub-servient to each other. And all the plants growing upon the surface of the earth, like that which I have brought here to serve as an illus-tration, absorb carbon; these leaves are taking up their carbon from the atmosphere to which we have given it in the form of carbonic acid, and they are growing and prospering. Give them a pure air like ours, and they could not live in it; give them carbon with other matters, and they live and rejoice. This piece of wood gets all its car-bon, as the trees and plants get theirs, from the atmosphere, which, as we have seen, carries away what is bad for us and at the same time good for them,—what is disease to the one being health to the other. So are we made dependent not merely upon our fellow-creatures, but upon our fellow-existers, all Nature being tied together by the laws that make one part conduce to the good of another.

There is another little point which I must mention before we draw to a close—a point which concerns the whole of these opera-tions, and most curious and beautiful it is to see it clustering upon

and associated with the bodies that concern us—oxygen, hydrogen, and carbon, in different states of their existence. I showed you
just now some powdered lead, which I set burning ([19]); and you
saw that the moment the fuel was brought to the air it acted, even
before it got out of the bottle,—the moment the air crept in it acted.
Now, there is a case of chemical affinity by which all our opera-
tions proceed. When we breathe, the same operation is going on
within us. When we burn a candle, the attraction of the different
parts one to the other is going on. Here it is going on in this case of
the lead, and it is a beautiful instance of chemical affinity. If the
products of combustion rose off from the surface, the lead would
take fire, and go on burning to the end; but you remember that we
have this difference between charcoal and lead—that, while the
lead can start into action at once if there be access of air to it, the
carbon will remain days, weeks, months, or years. The manuscripts
of Herculaneum were written with carbonaceous ink, and there
they have been for 1800 years or more, not having been at all
changed by the atmosphere, though coming in contact with it
under various circumstances. Now, what is the circumstance
which makes the lead and carbon differ in this respect? It is a striking thing to see that the matter which is appointed to serve the purpose of fuel *waits* in its action; it does not start off burning, like the
lead and many other things that I could show you, but which I
have not encumbered the table with; but it waits for action. This
waiting is a curious and wonderful thing. Candles—those Japanese
candles, for instance—do not start into action at once like the lead
or iron (for iron finely divided does the same thing as lead), but
there they wait for years, perhaps for ages, without undergoing
any alteration. I have here a supply of coal-gas. The jet is giving

---

[19] Water gas is formed by passing vapour of water over red-hot charcoal
or coke. It is a mixture of hydrogen and carbonic oxide, each of which is
an inflammable gas.

forth the gas, but you see it does not take fire—it comes out into the air, but it waits till it is hot enough before it burns. If I make it hot enough, it takes fire. If I blow it out, the gas that is issuing forth waits till the light is applied to it again. It is curious to see how different substances wait—how some will wait till the temperature is raised a little, and others till it is raised a good deal. I have here a little gunpowder and some gun-cotton; even these things differ in the conditions under which they will burn. The gunpowder is composed of carbon and other substances, making it highly combustible; and the gun-cotton is another combustible preparation. They are both waiting, but they will start into activity at different degrees of heat, or under different conditions. By applying a heated wire to them, we shall see which will start first [touching the gun-cotton with the hot iron]. You see the gun-cotton has gone off, but not even the hottest part of the wire is now hot enough to fire the gunpowder. How beautifully that shows you the difference in the degree in which bodies act in this way! In the one case the substance will wait any time until the associated bodies are made active by heat; but, in the other, as in the process of respiration, it waits no time. In the lungs, as soon as the air enters, it unites with the carbon; even in the lowest temperature which the body can bear short of being frozen, the action begins at once, producing the carbonic acid of respiration; and so all things go on fitly and properly. Thus you see the analogy between respiration and combustion is rendered still more beautiful and striking. Indeed, all I can say to you at the end of these lectures (for we must come to an end at one time or other) is to express a wish that you may, in your generation, be fit to compare to a candle; that you may, like it, shine as lights to those about you; that, in all your actions, you may justify the beauty of the taper by making your deeds honourable and effectual in the discharge of your duty to your fellow-men.

The Forms of Water in Clouds and Rivers, Ice and Glaciers

John Tyndall*

§ 1. *Clouds, Rains, and Rivers.*

1. EVERY occurrence in Nature is preceded by other occurrences which are its causes, and succeeded by others which are its effects. The human mind is not satisfied with observing and studying any natural occurrence alone, but takes pleasure in connecting every natural fact with what has gone before it, and with what is to come after it.

2. Thus, when we enter upon the study of rivers and glaciers, our interest will be greatly augmented by taking into account not only their actual appearances, but also their causes and effects.

3. Let us trace a river to its source. Beginning where it empties itself into the sea, and following it backwards, we find it from time to time joined by tributaries which swell its waters. The river of course becomes smaller as these tributaries are passed. It shrinks first to a brook, then to a stream; this again divides itself into a number of smaller streamlets, ending in mere threads of water. These constitute the source of the river, and are usually found among hills.

---

* The original version appeared in: John Tyndall, *The Forms of Water in Clouds & Rivers, Ice & Glaciers*, London, 1872, pp. 1–34.

4. Thus the Severn has its source in the Welsh Mountains; the Thames in the Cotswold Hills; the Danube in the hills of the Black Forest; the Rhine and the Rhone in the Alps; the Ganges in the Himalaya Mountains; the Euphrates near Mount Ararat; the Garonne in the Pyrenees; the Elbe in the Giant Mountains of Bohemia; the Missouri in the Rocky Mountains, and the Amazon in the Andes of Peru.

5. But it is quite plain that we have not yet reached the real beginning of the rivers. Whence do the earliest streams derive their water? A brief residence among the mountains would prove to you that they are fed by rains. In dry weather you would find the streams feeble, sometimes indeed quite dried up. In wet weather you would see them foaming torrents. In general these streams lose themselves as little threads of water upon the hill sides; but sometimes you may trace a river to a definite spring. The river Albula in Switzerland, for instance, rushes at its origin in considerable volume from a mountain side. But you very soon assure yourself that such springs are also fed by rain, which has percolated through the rocks or soil, and which, through some orifice that it has found or formed, comes to the light of day.

6. But we cannot end here. Whence comes the rain which forms the mountain streams? Observation enables you to answer the question. Rain does not come from a clear sky. It comes from clouds. But what are clouds? Is there nothing you are acquainted with which they resemble? You discover at once a likeness between them and the condensed steam of a locomotive. At every puff of the engine a cloud is projected into the air. Watch the cloud sharply: you notice that it first forms at a little distance from the top of the funnel. Give close attention and you will sometimes see a perfectly clear space between the funnel and the cloud. Through that clear space the thing which makes the cloud must pass. What, then, is this thing which at one moment is transparent and invisible, and at the next moment visible as a dense opaque cloud?

7. It is the *steam* or *vapour of water* from the boiler. Within the boiler this steam is transparent and invisible; but to keep it in this invisible state a heat would be required as great as that within the boiler. When the vapour mingles with the cold air above the hot funnel it ceases to be vapour. Every bit of steam shrinks, when chilled, to a much more minute particle of water. The liquid particles thus produced form a kind of *water-dust* of exceeding fineness, which floats in the air, and is called *a cloud.*

8. Watch the cloud-banner from the funnel of a running locomotive; you see it growing gradually less dense. It finally melts away altogether, and if you continue your observations you will not fail to notice that the speed of its disappearance depends upon the character of the day. In humid weather the cloud hangs long and lazily in the air; in dry weather it is rapidly licked up. What has become of it? It has been reconverted into true invisible vapour.

9. The *drier* the air, and the *hotter* the air, the greater is the amount of cloud which can be thus dissolved in it. When the cloud first forms, its quantity is far greater than the air is able to maintain in an invisible state. But as the cloud mixes gradually with a larger mass of air it is more and more dissolved, and finally passes altogether from the condition of a finely-divided liquid into that of transparent vapour or gas.

10. Make the lid of a kettle air-tight, and permit the steam to issue from the pipe; a cloud is precipitated in all respects similar to that issuing from the funnel of the locomotive.

11. Permit the steam as it issues from the pipe to pass through the flame of a spirit-lamp, the cloud is instantly dissolved by the heat, and is not again precipitated. With a special boiler and a special nozzle the experiment may be made more striking, but not more instructive, than with the kettle.

12. Look to your bedroom windows when the weather is very cold outside; they sometimes stream with water derived from the

condensation of the aqueous vapour from your own lungs. The windows of railway carriages in winter show this condensation in a striking manner. Pour cold water into a dry drinking-glass on a summer's day: the outside surface of the glass becomes instantly dimmed by the precipitation of moisture. On a warm day you notice no vapour in front of your mouth, but on a cold day you form there a little cloud derived from the condensation of the aqueous vapour from the lungs.

13. You may notice in a ball-room that as long as the door and windows are kept closed, and the room remains hot, the air remains clear; but when the doors or windows are opened a dimness is visible, caused by the precipitation to fog of the aqueous vapour of the ball-room. If the weather be intensely cold the entrance of fresh air may even cause *snow* to fall. This has been observed in Russian ball-rooms; and also in the subterranean stables at Erzeroom, when the doors are opened and the cold morning air is permitted to enter.

14. Even on the driest day this vapour is never absent from our atmosphere. The vapour diffused through the air of this room may be congealed to hoar frost in your presence. This is done by filling a vessel with a mixture of pounded ice and salt, which is colder than the ice itself, and which, therefore, condenses and freezes the aqueous vapour. The surface of the vessel is finally coated with a frozen fur, so thick that it may be scraped away and formed into a snow-ball.

15. To produce the cloud, in the case of the locomotive and the kettle, *heat* is necessary. By heating the water we first convert it into steam, and then by chilling the steam we convert it into cloud. Is there any fire in nature which produces the clouds of our atmosphere? There is: the fire of the sun.

16. Thus, by tracing backward, without any break in the chain of occurrences, our river from its end to its real beginnings, we come at length to the sun.

§ 2.

17. There are, however, rivers which have sources somewhat different from those just mentioned. They do not begin by driblets on a hill side, nor can they be traced to a spring. Go, for example, to the mouth of the river Rhone, and trace it backwards to Lyons, where it turns to the east. Bending round by Chambery, you come at length to the Lake of Geneva, from which the river rushes, and which you might be disposed to regard as the source of the Rhone. But go to the head of the lake, and you find that the Rhone there enters it, that the lake is in fact a kind of expansion of the river. Follow this upwards; you find it joined by smaller rivers from the mountains right and left. Pass these, and push your journey higher still. You come at length to a huge mass of ice—the end of a glacier—which fills the Rhone valley, and from the bottom of the glacier the river rushes. In the glacier of the Rhone you thus find the source of the river Rhone.

18. But again we have not reached the real beginning of the river. You soon convince yourself that this earliest water of the Rhone is produced by the melting of the ice. You get upon the glacier and walk upwards along it. After a time the ice disappears and you come upon snow. If you are a competent mountaineer you may go to the very top of this great snow-field, and if you cross the top and descend at the other side you finally quit the snow, and get upon another glacier called the Trift, from the end of which rushes a river smaller than the Rhone.

19. You soon learn that the mountain snow feeds the glacier. By some means or other the snow is converted into ice. But whence comes the snow? Like the rain, it comes from the clouds, which, as before, can be traced to vapour raised by the sun. Without solar fire we could have no atmospheric vapour, without vapour no clouds, without clouds no snow, and without snow no glaciers. Curious then as the conclusion may be, the cold ice of the Alps has its origin in the heat of the sun.

§ 3. *The Waves of Light.*

20. But what is the sun? We know its size and its weight. We also know that it is a globe of fire far hotter than any fire upon earth. But we now enter upon another enquiry. We have to learn definitely what is the meaning of solar light and solar heat; in what way they make themselves known to our senses; by what means they get from the sun to the earth, and how, when there, they produce the clouds of our atmosphere, and thus originate our rivers and our glaciers.

21. If in a dark room you close your eyes and press the eyelid with your finger-nail, a circle of light will be seen opposite to the point pressed, while a sharp blow upon the eye produces the impression of a flash of light. There is a nerve specially devoted to the purposes of vision which comes from the brain to the back of the eye, and there divides into fine filaments, which are woven together to a kind of screen called the *retina*. The retina can be excited in various ways so as to produce the consciousness of light; it may, as we have seen, be excited by the rude mechanical action of a blow imparted to the eye.

22. There is no spontaneous creation of light by the healthy eye. To excite vision the retina must be affected by something coming from without. What is that something? In some way or other luminous bodies have the power of affecting the retina—but *how?*

23. It was long supposed that from such bodies issued, with inconceivable rapidity, an inconceivably fine matter, which flew through space, passed through the pores supposed to exist in the humours of the eye, reached the retina behind, and by their shock against the retina, aroused the sensation of light.

24. This theory, which was supported by the greatest men, among others by Sir Isaac Newton, was found competent to explain a great number of the phenomena of light, but it was not found competent to explain *all* the phenomena. As the skill and knowledge of experimenters increased, large classes of facts were

revealed which could only be explained by assuming that light was produced, not by a fine matter flying through space and hitting the retina, but by the shock of minute *waves* against the retina.

25. Dip your finger into a basin of water, and cause it to quiver rapidly to and fro. From the point of disturbance issue small ripples which are carried forward by the water, and which finally strike the basin. Here, in the vibrating finger, you have a source of agitation; in the water you have a vehicle through which the finger's motion is transmitted, and you have finally the side of the basin which receives the shock of the little waves.

26. In like manner, according to the *wave theory* of light, you have a source of agitation in the vibrating atoms, or smallest particles, of the luminous body; you have a vehicle of transmission in a substance which is supposed to fill all space, and to be diffused through the humours of the eye; and finally, you have the retina, which receives the successive shocks of the waves. These shocks are supposed to produce the sensation of light.

27. We are here dealing, for the most part, with suppositions and assumptions merely. We have never seen the atoms of a luminous body, nor their motions. We have never seen the medium which transmits their motions, nor the waves of that medium. How, then, do we come to assume their existence?

28. Before such an idea could have taken any real root in the human mind, it must have been well disciplined and prepared by observations and calculations of ordinary wave-motion. It was necessary to know how both water-waves and sound-waves are formed and propagated. It was above all things necessary to know how waves, passing through the same medium, act upon each other. Thus disciplined, the mind was prepared to detect any resemblance presenting itself between the action of light and that of waves. Great classes of optical phenomena accordingly appeared which could be accounted for in the most complete and satisfactory manner by assuming them to be produced by waves, and which

could not be otherwise accounted for. It is because of its competence to explain all the phenomena of light that the wave theory now receives universal acceptance on the part of scientific men.

Let me use an illustration. We infer from the flint implements recently found in such profusion all over England and in other countries, that they were produced by men, and also that the Pyramids of Egypt were built by men, because, as far as our experience goes, nothing but men could form such implements or build such Pyramids. In like manner, we infer from the phenomena of light the agency of waves, because, as far as our experience goes, no other agency could produce the phenomena.

§ 4. *The Waves of Heat which produce the Vapour of our Atmosphere and melt our Glaciers.*

29. Thus, in a general way, I have given you the conception and the grounds of the conception, which regards light as the product of wave-motion; but we must go farther than this, and follow the conception into some of its details. We have all seen the waves of water, and we know they are of different sizes—different in length and different in height. When, therefore, you are told that the atoms of the sun, and of almost all other luminous bodies, vibrate at different rates, and produce waves of different sizes, your experience of water-waves will enable you to form a tolerably clear notion of what is meant.

30. As observed above, we have never seen the light-waves, but we judge of their presence, their position, and their magnitude, by their effects. Their lengths have been thus determined, and found to vary from about $\frac{1}{30000}$th to $\frac{1}{60000}$th of an inch.

31. But besides those which produce light, the sun sends forth incessantly a multitude of waves which produce no light. The largest waves which the sun sends forth are of this non-luminous character, though they possess the highest heating power.

32. A common sunbeam contains waves of all kinds, but it is possible to *sift* or *filter* the beam so as to intercept all its light, and to allow its obscure heat to pass unimpeded. For substances have been discovered which, while intensely opaque to the light-waves, are almost perfectly transparent to the others. On the other hand, it is possible, by the choice of proper substances, to intercept in a great degree the pure heat-waves, and to allow the pure light-waves free transmission. This last separation is, however, not so perfect as the first.

33. We shall learn presently how to detach the one class of waves from the other class, and to prove that waves competent to light a fire, fuse metal, or burn the hand like a hot solid, may exist in a perfectly dark place.

34. Supposing, then, that we withdraw, in the first instance the large heat-waves, and allow the light-waves alone to pass. These may be concentrated by suitable lenses and sent into water without sensibly warming it. Let the light-waves now be withdrawn, and the larger heat-waves concentrated in the same manner; they may be caused to boil the water almost instantaneously.

35. This is the point to which I wished to lead you, and which without due preparation could not be understood. You now perceive the important part played by these large darkness-waves, if I may use the term, in the work of evaporation. When they plunge into seas, lakes, and rivers, they are intercepted close to the surface, and they heat the water at the surface, thus causing it to evaporate; the light-waves at the same time entering to great depths without sensibly heating the water through which they pass. Not only, therefore, is it the sun's fire which produces evaporation, but a particular constituent of that fire, the existence of which you probably were not aware of.

36. Further, it is these selfsame lightless waves which, falling upon the glaciers of the Alps, melt the ice and produce all the rivers flowing from the glaciers; for I shall prove to you presently that the

light-waves, even when concentrated to the uttermost, are unable to melt the most delicate hoar-frost; much less would they be able to produce the copious liquefaction observed upon the glaciers.

37. These large lightless waves of the sun, as well as the heat-waves issuing from non-luminous hot bodies, are frequently called obscure or invisible heat.

We have here an example of the manner in which phenomena, apparently remote, are connected together in this wonderful system of things that we call Nature. You cannot study a snow-flake profoundly without being led back by it step by step to the constitution of the sun. It is thus throughout Nature. All its parts are interdependent, and the study of any one part *completely* would really involve the study of all.

§ 5. *Experiments to prove the foregoing Statements.*

38. Heat issuing from any source not visibly red cannot be concentrated so as to produce the intense effects just referred to. To produce these it is necessary to employ the obscure heat of a body raised to the highest possible state of incandescence. The sun is such a body, and its dark heat is therefore suitable for experiments of this nature.

39. But in the atmosphere of London, and for experiments such as ours, the heat-waves emitted by coke raised to intense whiteness by a current of electricity are much more manageable than the sun's waves. The electric light has also the advantage that its dark radiation embraces a larger proportion of the total radiation than the dark heat of the sun. In fact, the force or energy, if I may use the term, of the dark waves of the electric light is fully seven times that of its light-waves. The electric light, therefore, shall be employed in our experimental demonstrations.

40. From this source a powerful beam is sent through the room, revealing its track by the motes floating in the air of the room; for

were the motes entirely absent the beam would be unseen. It falls upon a concave mirror (a glass one silvered behind will answer) and is gathered up by the mirror into a cone of reflected rays; the luminous apex of the cone, which is the *focus* of the mirror, being about fifteen inches distant from its reflecting surface. Let us mark the focus accurately by a pointer.

41. And now let us place in the path of the beam a substance perfectly opaque to light. This substance is iodine dissolved in a liquid called bisulphide of carbon. The light at the focus instantly vanishes when the dark solution is introduced. But the solution is intensely transparent to the dark waves, and a focus of such waves remains in the air of the room after the light has been abolished. You may feel the heat of these waves with your hand; you may let them fall upon a thermometer, and thus prove their presence; or, best of all, you may cause them to produce a current of electricity, which deflects a large magnetic needle. The magnitude of the deflection is a measure of the heat.

42. Our object now is, by the use of a more powerful lamp, and a better mirror (one silvered in front and with a shorter focal distance), to intensify the action here rendered so sensible. As before, the focus is rendered strikingly visible by the intense illumination of the dust particles. We will first filter the beam so as to intercept its dark waves, and then permit the purely luminous waves to exert their utmost power on a small bundle of gun-cotton placed at the focus.

43. No effect whatever is produced. The gun-cotton might remain there for a week without ignition. Let us now permit the unfiltered beam to act upon the cotton. It is instantly dissipated in an explosive flash. This experiment proves that the light-waves are incompetent to explode the cotton, while the waves of the full beam are competent to do so; hence we may conclude that the dark waves are the real agents in the explosion.

44. But this conclusion would be only probable; for it might be urged that the *mixture* of the dark waves and the light-waves is

necessary to produce the result. Let us then, by means of our opaque solution, isolate our dark waves and converge them on the cotton. It explodes as before.

45. Hence it is the dark waves, and they only, that are concerned in the ignition of the cotton.

46. At the same dark focus sheets of platinum are raised to vivid redness; zinc is burnt up; paper instantly blazes; magnesium wire is ignited; charcoal within a receiver containing oxygen is set burning; a diamond similarly placed is caused to glow like a star, being afterwards gradually dissipated. And all this while the *air* at the focus remains as cool as in any other part of the room.

47. To obtain the light-waves we employ a clear solution of alum in water; to obtain the dark waves we employ the solution of iodine above referred to. But as before stated (32), the alum is not so perfect a filter as the iodine; for it transmits a portion of the obscure heat.

48. Though the light-waves here prove their incompetence to ignite gun-cotton, they are able to burn up black paper; or, indeed, to explode the cotton when it is blackened. The white cotton does not absorb the light, and without absorption we have no heating. The blackened cotton absorbs, is heated, and explodes.

49. Instead of a solution of alum, we will employ for our next experiment a cell of pure water, through which the light passes without sensible absorption. At the focus is placed a test-tube also containing water, the full force of the light being concentrated upon it. The water is not sensibly warmed by the concentrated waves. We now remove the cell of water; no change is visible in the beam, but the water contained in the test-tube now boils.

50. The light-waves being thus proved ineffectual, and the full beam effectual, we may infer that it is the dark waves that do the work of heating. But we clench our inference by employing our opaque iodine filter. Placing it on the path of the beam, the light is entirely stopped, but the water boils exactly as it did when the full beam fell upon it.

51. The truth of the statement made in paragraph (34) is thus demonstrated.

52. And now with regard to the melting of ice. On the surface of a flask containing a freezing mixture we obtain a thick fur of hoar-frost. Sending the beam through a water-cell, its luminous waves are concentrated upon the surface of the flask. Not a spicula of the frost is dissolved. We now remove the water-cell, and in a moment a patch of the frozen fur as large as half-a-crown is melted. Hence, inasmuch as the full beam produces this effect, and the luminous part of the beam does not produce it, we fix upon the dark portion the melting of the frost.

53. As before, we clench this inference by concentrating the dark waves alone upon the flask. The frost is dissipated exactly as it was by the full beam.

54. These effects are rendered strikingly visible by darkening with ink the freezing mixture within the flask. When the hoar frost is removed, the blackness of the surface from which it had been melted comes out in strong contrast with the adjacent snowy whiteness. When the flask itself, instead of the freezing mixture, is blackened, the purely luminous waves being absorbed by the glass, warm it; the glass reacts upon the frost, and melts it. Hence the wisdom of darkening, instead of the flask itself, the mixture within the flask.

55. This experiment proves to demonstration the statement in paragraph (36): that it is the dark waves of the sun that melt the mountain snow and ice, and originate all the rivers derived from glaciers.

There are writers who seem to regard science as an aggregate of facts, and hence doubt its efficacy as an exercise of the reasoning powers. But all that I have here taught you is the result of reason, taking its stand, however, upon the sure basis of observation and experiment. And this is the spirit in which our further studies are to be pursued.

§ 6. *Oceanic Distillation.*

56. The sun, you know, is never exactly overhead in England. But at the equator, and within certain limits north and south of it, the sun at certain periods of the year is directly overhead at noon. These limits are called the Tropics of Cancer and of Capricorn. Upon the belt comprised between these two circles the sun's rays fall with their mightiest power; for here they shoot directly downwards, and heat both earth and sea more than when they strike slantingly.

57. When the vertical sunbeams strike the land they heat it, and the air in contact with the hot soil becomes heated in turn. But when heated the air expands, and when it expands it becomes lighter. This lighter air rises, like wood plunged into water, through the heavier air overhead.

58. When the sunbeams fall upon the sea the water is warmed, though not so much as the land. The warmed water expands, becomes thereby lighter, and therefore continues to float upon the top. This upper layer of water warms to some extent the air in contact with it, but it also sends up a quantity of aqueous vapour, which being far lighter than air, helps the latter to rise. Thus both from the land and from the sea we have ascending currents established by the action of the sun.

59. When they reach a certain elevation in the atmosphere, these currents divide and flow, part towards the north and part towards the south; while from the north and the south a flow of heavier and colder air sets in to supply the place of the ascending warm air.

60. Incessant circulation is thus established in the atmosphere. The equatorial air and vapour flow above towards the north and south poles, while the polar air flows below towards the equator. The two currents of air thus established are called the upper and the lower trade winds.

61. But before the air returns from the poles great changes have occurred. For the air as it quitted the equatorial regions was laden with aqueous vapour, which could not subsist in the cold polar

regions. It is there precipitated, falling sometimes as rain, or more commonly as snow. The land near the pole is covered with this snow, which gives birth to vast glaciers in a manner hereafter to be explained.

62. It is necessary that you should have a perfectly clear view of this process, for great mistakes have been made regarding the manner in which glaciers are related to the heat of the sun.

63. It was supposed that if the sun's heat were diminished, greater glaciers than those now existing would be produced. But the lessening of the sun's heat would infallibly diminish the quantity of aqueous vapour, and thus cut off the glaciers at their source. A brief illustration will complete your knowledge here.

64. In the process of ordinary distillation, the liquid to be distilled is heated and converted into vapour in one vessel, and chilled and reconverted into liquid in another. What has just been stated renders it plain that the earth and its atmosphere constitute a vast distilling apparatus in which the equatorial ocean plays the part of the boiler, and the chill regions of the poles the part of the condenser. In this process of distillation *heat* plays quite as necessary a part as *cold*, and before Bishop Heber could speak of 'Greenland's icy mountains,' the equatorial ocean had to be warmed by the sun. We shall have more to say upon this question afterwards.

ILLUSTRATIVE EXPERIMENTS.

65. I have said that when heated, air expands. If you wish to verify this for yourself, proceed thus. Take an empty flask, stop it by a cork; pass through the cork a narrow glass tube. By heating the tube in a spirit-lamp you can bend it downwards, so that when the flask is standing upright the open end of the narrow tube may dip into water. Now cause the flame of your spirit-lamp to play against the flask. The flame heats the glass, the glass heats the air; the air expands, is driven through the narrow tube, and issues in a storm of bubbles from the water.

66. Were the heated air unconfined, it would rise in the heavier cold air. Allow a sunbeam or any other intense light to fall upon a white wall or screen in a dark room. Bring a heated poker, a candle, or a gas flame underneath the beam. An ascending current rises from the heated body through the beam, and the action of the air upon the light is such as to render the wreathing and waving of the current strikingly visible upon the screen. When the air is hot enough, and therefore light enough, if entrapped in a paper bag it carries the bag upwards, and you have the Fire-balloon.

67. Fold two sheets of paper into two cones and suspend them with their closed points upwards from the end of a delicate balance. See that the cones balance each other. Then place for a moment the flame of a spirit-lamp beneath the open base of one of them; the hot air ascends from the lamp and instantly tosses upwards the cone above it.

68. Into an inverted glass shade introduce a little smoke. Let the air come to rest, and then simply place your hand at the open mouth of the shade. Mimic hurricanes are produced by the air warmed by the hand, which are strikingly visible when the smoke is illuminated by a strong light.

69. The heating of the tropical air by the sun is *indirect*. The solar beams have scarcely any power to heat the air through which they pass; but they heat the land and ocean, and these communicate their heat to the air in contact with them. The air and vapour start upwards charged with the heat thus communicated.

§ 7. *Tropical Rains.*

70. But long before the air and vapour from the equator reach the poles, precipitation occurs. Wherever a humid warm wind mixes with a cold dry one, rain falls. Indeed the heaviest rains occur at those places where the sun is vertically overhead. We must enquire a little more closely into their origin.

71. Fill a bladder about two-thirds full of air at the sea level, and take it to the summit of Mont Blanc. As you ascend, the bladder becomes more and more distended; at the top of the mountain it is fully distended, and has evidently to bear a pressure from within. Returning to the sea level you find that the tightness disappears, the bladder finally appearing as flaccid as at first.

72. The reason is plain. At the sea level the air within the bladder has to bear the pressure of the whole atmosphere, being thereby squeezed into a comparatively small volume. In ascending the mountain, you leave more and more of the atmosphere behind; the pressure becomes less and less, and by its expansive force the air within the bladder swells as the outside pressure is diminished. At the top of the mountain the expansion is quite sufficient to render the bladder tight, the pressure within being then actually greater than the pressure without. By means of an air-pump we can show the expansion of a balloon partly filled with air, when the external pressure has been in part removed.

73. But why do I dwell upon this? Simply to make plain to you that the *unconfined air*, heated at the earth's surface, and ascending by its lightness, must expand more and more the higher it rises in the atmosphere.

74. And now I have to introduce to you a new fact, towards the statement of which I have been working for some time. It is this:— *The ascending air is chilled by its expansion.* Indeed this chilling is one source of the coldness of the higher atmospheric regions. And now fix your eye upon those mixed currents of air and aqueous vapour which rise from the warm tropical ocean. They start with plenty of heat to preserve the vapour as vapour; but as they rise they come into regions already chilled, and they are still further chilled by their own expansion. The consequence might be foreseen. The load of vapour is in great part precipitated, dense clouds are formed, their particles coalesce to rain-drops, which descend daily in gushes so profuse that the word 'torrential' is

used to express the copiousness of the rain-fall. I could show you this chilling by expansion, and also the consequent precipitation of clouds.

75. Thus long before the air from the equator reaches the poles its vapour is in great part removed from it, having redescended to the earth as rain. Still a good quantity of the vapour is carried forward, which yields hail, rain, and snow in northern and southern lands.

ILLUSTRATIVE EXPERIMENTS.

76. I have said that the air is chilled during its expansion. Prove this, if you like, thus. With a condensing syringe, you can force air into an iron box furnished with a stopcock, to which the syringe is screwed. Do so till the density of the air within the box is doubled or trebled. Immediately after this condensation, both the box and the air within it are *warm,* and can be proved to be so by a proper thermometer. Simply turn the cock and allow the compressed air to stream into the atmosphere. The current, if allowed to strike a thermometer, visibly chills it; and with other instruments the chill may be made more evident still. Even the hand feels the chill of the expanding air.

77. Throw a strong light, a concentrated sunbeam for example, across the issuing current; if the compressed air be ordinary humid air, you see the precipitation of a little cloud by the chill accompanying the expansion. This cloud-formation may, however, be better illustrated in the following way:—

78. In a darkened room send a strong beam of light through a glass tube three feet long and three inches wide, stopped at its ends by glass plates. Connect the tube by means of a stopcock with a vessel of about one-fourth its capacity, from which the air has been removed by an air-pump. The exhausted cylinder of the pump itself will answer capitally. Fill the glass tube with humid air; then simply turn on the stopcock which connects it with the exhausted vessel. Having more room the air expands, cold accompanies the expansion, and, as a consequence, a dense and brilliant cloud

immediately fills the tube. If the experiment be made for yourself alone you may see the cloud in ordinary daylight; indeed, the brisk exhaustion of any receiver filled with humid air is known to produce this condensation.

79. Other vapours than that of water may be thus precipitated, some of them yielding clouds of intense brilliancy, and displaying iridescences, such as are sometimes, but not frequently, seen in the clouds floating over the Alps.

80. In science what is true for the small is true for the large. Thus by combining the conditions observed on a large scale in nature we obtain on a small scale the phenomena of atmospheric clouds.

§ 8. *Mountain Condensers.*

81. To complete our view of the process of atmospheric precipitation we must take into account the action of mountains. Imagine a south-west wind blowing across the Atlantic towards Ireland. In its passage it charges itself with aqueous vapour. In the south of Ireland it encounters the mountains of Kerry: the highest of these is Magillicuddy's Reeks, near Killarney. Now the lowest stratum of this Atlantic wind is that which is most fully charged with vapour. When it encounters the base of the Kerry mountains it is tilted up and flows bodily over them. Its load of vapour is therefore carried to a height, it expands on reaching the height, it is chilled in consequence of the expansion, and comes down in copious showers of rain. From this, in fact, arises the luxuriant vegetation of Killarney; to this, indeed, the lakes owe their water supply. The cold crests of the mountains also aid in the work of condensation.

82. Note the consequence. There is a town called Cahirciveen to the south-west of Magillicuddy's Reeks, at which observations of the rainfall have been made, and a good distance farther to the north-east, right in the course of the south-west wind, there is another town, called Portarlington, at which observations of rainfall have also been made. But before the wind reaches the latter station

it has passed over the mountains of Kerry and left a great portion of its moisture behind it. What is the result? At Cahirciveen, as shown by Dr. Lloyd, the rainfall amounts to 59 inches in a year, while at Portarlington it is only 21 inches.

83. Again, you may sometimes descend from the Alps when the fall of rain and snow is heavy and incessant, into Italy, and find the sky over the plains of Lombardy blue and cloudless, the wind at the same time *blowing over the plain towards the Alps.* Below the wind is hot enough to keep its vapour in a perfectly transparent state; but it meets the mountains, is tilted up, expanded, and chilled. The cold of the higher summits also helps the chill. The consequence is that the vapour is precipitated as rain or snow, thus producing bad weather upon the heights, while the plains below, flooded with the same air, enjoy the aspect of the unclouded summer sun. Clouds blowing *from* the Alps are also sometimes dissolved over the plains of Lombardy.

84. In connection with the formation of clouds by mountains, one particularly instructive effect may be here noticed. You frequently see a streamer of cloud many hundred yards in length drawn out from an Alpine peak. Its steadiness appears perfect, though a strong wind may be blowing at the same time over the mountain head. Why is the cloud not blown away? It *is* blown away; its permanence is only apparent. At one end it is incessantly dissolved, at the other end it is incessantly renewed: supply and consumption being thus equalized, the cloud appears as changeless as the mountain to which it seems to cling. When the red sun of the evening shines upon these cloud-streamers they resemble vast torches with their flames blown through the air.

§ 9. *Architecture of Snow.*

85. We now resemble persons who have climbed a difficult peak, and thereby earned the enjoyment of a wide prospect. Having

made ourselves masters of the conditions necessary to the production of mountain snow, we are able to take a comprehensive and intelligent view of the phenomena of glaciers.

86. A few words are still necessary as to the formation of snow. The molecules and atoms of all substances, when allowed free play, build themselves into definite and, for the most part, beautiful forms called crystals. Iron, copper, gold, silver, lead, sulphur, when melted and permitted to cool gradually, all show this crystallizing power. The metal bismuth shows it in a particularly striking manner, and when properly fused and solidified, self-built crystals of great size and beauty are formed of this metal.

87. If you dissolve saltpetre in water, and allow the solution to evaporate slowly, you may obtain large crystals, for no portion of the salt is converted into vapour. The water of our atmosphere is fresh though it is derived from the salt sea. Sugar dissolved in water, and permitted to evaporate, yields crystals of sugar-candy. Alum readily crystallizes in the same way. Flints dissolved, as they sometimes are in nature, and permitted to crystallize, yield the prisms and pyramids of rock crystal. Chalk dissolved and crystallized yields Iceland spar. The diamond is crystallized carbon. All our precious stones, the ruby, sapphire, beryl, topaz, emerald, are all examples of this crystallizing power.

88. You have heard of the force of gravitation, and you know that it consists of an attraction of every particle of matter for every other particle. You know that planets and moons are held in their orbits by this attraction. But gravitation is a very simple affair compared to the force, or rather forces, of crystallization. For here the ultimate particles of matter, inconceivably small as they are, show themselves possessed of attractive and repellent poles, by the mutual action of which the shape and structure of the crystal are determined. In the solid condition the attracting poles are rigidly locked together; but if sufficient heat be applied the bond of union is dissolved, and in the state of fusion the poles are pushed so far asunder as to be practically

out of each other's range. The natural tendency of the molecules to build themselves together is thus neutralized.

89. This is the case with water, which as a liquid is to all appearance formless. When sufficiently cooled the molecules are brought within the play of the crystallizing force, and they then arrange themselves in forms of indescribable beauty. When snow is produced in calm air, the icy particles build themselves into beautiful stellar shapes, each star possessing six rays. There is no deviation from this type, though in other respects the appearances of the snow-stars are infinitely various. In the polar regions these exquisite forms were observed by Dr. Scoresby, who gave numerous drawings of them. I have observed them in mid-winter filling the air, and loading the slopes of the Alps. But in England they are also to be seen, and no words of mine could convey so vivid an impression of their beauty as the annexed drawings of a few of them, executed at Greenwich by Mr. Glaisher.

90. It is worth pausing to think what wonderful work is going on in the atmosphere during the formation and descent of every snow-shower: what building power is brought into play! and how imperfect seem the productions of human minds and hands when compared with those formed by the blind forces of nature!

91. But who ventures to call the forces of nature blind? In reality, when we speak thus we are describing our own condition. The blindness is ours; and what we really ought to say, and to confess, is that our powers are absolutely unable to comprehend either the origin or the end of the operations of nature.

92. But while we thus acknowledge our limits, there is also reason for wonder at the extent to which science has mastered the system of nature. From age to age, and from generation to generation, fact has been added to fact, and law to law, the true method and order of the Universe being thereby more and more revealed. In doing this science has encountered and overthrown various

forms of superstition and deceit, of credulity and imposture. But the world continually produces weak persons and wicked persons; and as long as they continue to exist side by side, as they do in this our day, very debasing beliefs will also continue to infest the world.

§ 10. *Atomic Poles.*

93. 'What did I mean when, a few moments ago (88), I spoke of attracting and repellent poles?' Let me try to answer this question. You know that astronomers and geographers speak of the earth's poles, and you have also heard of magnetic poles, the poles of a magnet being the points at which the attraction and repulsion of the magnet are as it were concentrated.

94. Every magnet possesses two such poles; and if iron filings be scattered over a magnet, each particle becomes also endowed with two poles. Suppose such particles devoid of weight and floating in our atmosphere, what must occur when they come near each other? Manifestly the repellent poles will retreat from each other, while the attractive poles will approach and finally lock themselves together. And supposing the particles, instead of a single pair, to possess several pairs of poles arranged at definite points over their surfaces; you can then picture them, in obedience to their mutual attractions and repulsions, building themselves together to form masses of definite shape and structure.

95. Imagine the molecules of water in calm cold air to be gifted with poles of this description, which compel the particles to lay themselves together in a definite order, and you have before your mind's eye the unseen architecture which finally produces the visible and beautiful crystals of the snow. Thus our first notions and conceptions of poles are obtained from the sight of our eyes in looking at the effects of magnetism; and we then transfer these

SNOW CRYSTALS.

D

notions and conceptions to particles which no eye has ever seen. The power by which we thus picture to ourselves effects beyond the range of the senses is what philosophers call the Imagination, and in the effort of the mind to seize upon the unseen architecture of crystals, we have an example of the 'scientific use' of this faculty. Without imagination we might have *critical* power, but not *creative* power in science.

Lessons in Electricity

John Tyndall*

§ 29. *Physiological Effects of the Electric Discharge.*

The physiological effect of the electric shock has been studied in various ways. Graham caused a number of persons to lay hold of the same metal plate, which was connected with the outer coating of a charged Leyden jar, and also to lay hold of a rod by which the jar was discharged. The shock divided itself equally among them.

The Abbé Nollet formed a line of one hundred and eighty guardsmen, and sent the discharge through them all. He also killed sparrows and fishes by the shock. The analogy of these effects with those produced by thunder and lightning could not escape attention, nor fail to stimulate enquiry.

Indeed, as experimental knowledge increased, men's thoughts became more definite and exact as regards the relation of electrical effects to thunder and lightning. The Abbé Nollet thus quaintly expresses himself: 'If any one should take upon him to prove, from a well-connected comparison of phenomena, that thunder is, in the hands of Nature, what electricity is in ours, and that the wonders which we now exhibit at our pleasure are little imitations of those

---

* The original version appeared in: John Tyndall, *Lessons in Electricity at the Royal Institution 1875–6*, London, 1876, pp. 97–111.

great effects which frighten us; I avow that this idea, if it was well supported, would give me a great deal of pleasure.' He then points out the analogies between both, and continues thus: 'All those points of analogy, which I have been some time meditating, begin to make me believe that one might, by taking electricity as the model, form to one's self, in relation to thunder and lightning, more perfect and more probable ideas than what have been offered hitherto.'[1]

These views were prevalent at the time now referred to, and out of them grew the experimental proof by the great physical philosopher, Franklin, of the substantial identity of the lightning flash and the electric spark.

Franklin was twice struck senseless by the electric shock. He afterwards sent the discharge of two large jars through six robust men; they fell to the ground and got up again without knowing what had happened; they neither heard nor felt the discharge. Priestley, who made many valuable contributions to electricity, received the charge of two jars, but did not find it painful.

This experience agrees with mine. Some time ago I stood in this room with a charged battery of fifteen large Leyden jars beside me. Through some awkwardness on my part I touched the wire leading from the battery, and the discharge went through me. For a sensible interval life was absolutely blotted out, but there was no trace of pain. After a little time consciousness returned; I saw confusedly both the audience and the apparatus, and concluded from this, and from my own condition, that I had received the discharge. To prevent the audience from being alarmed, I made the remark that it had often been my desire to receive such a shock accidentally, and that my wish had at length been fulfilled. But though the *intellectual* consciousness of my position returned with exceeding rapidity, it was not so with the *optical* consciousness. For, while making

---

[1] Priestley's 'History of Electricity,' pp. 151–52.

the foregoing remark, my body presented to my eyes the appearance of a number of separate pieces. The arms, for example, were detached from the trunk and suspended in the air. In fact, memory, and the power of reasoning, appeared to be complete, long before the restoration of the optic nerve to healthy action.

This may be regarded as an experimental proof that people killed by lightning suffer no pain.

§ 30. *Atmospheric Electricity.*

The air at all times can be proved to be a reservoir of electricity, which undergoes periodic variation. We have seen that ingenious men began soon to suspect a common origin for the crackling and light of the electric spark, and thunder and lightning. The greatest investigator in this field is the celebrated Dr. Franklin. He made an exhaustive comparison of the effects of electricity and those of lightning. The lightning flash he saw was of the same shape as the elongated electric spark; like electricity, lightning strikes pointed objects in preference to others; lightning pursues the path of least resistance; it burns, dissolves metals, rends bodies asunder, and strikes men blind. Franklin imitated all these effects, striking a pigeon blind, and killing a hen and turkey by the electrical discharge. I place before you in fig. 54, with a view to its comparison with a discharge of forked lightning, the long spark obtained from

Fig. 54.

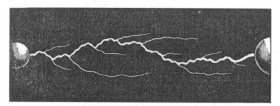

an effective ebonite machine, furnished with a conductor of a special construction, which favours length of spark.

Having completely satisfied his mind by this comparison of the identity of both agents, Franklin proposed to draw electricity from the clouds by a pointed rod erected on a high tower. But before the tower could be built he succeeded in his object by means of a kite with a pointed wire attached to it. The electricity descended by the hempen string which held the kite, to a key at the end of it, the key being separated from the observer by a silken string held in the hand. Franklin thus obtained sparks, and charged a Leyden phial with atmospheric electricity.

But, spurred by Franklin's researches, an observer in France had previously proved the electrical character of lightning. A translation of Franklin's writings on the subject fell into the hands of the naturalist Buffon, who requested his friend D'Alibard to revise the translation. D'Alibard was thus induced to erect an iron rod 40 feet long, supported by silk strings, and ending in a sentry-box. It was watched by an old dragoon named Coiffier, who on the 10th of May, 1752, heard a clap of thunder, and immediately afterwards drew sparks from the end of the iron rod.

The danger of experiments with metal rods was soon illustrated. Professor Richmann of St. Petersburg had a rod raised three or four feet above the tiles of his house. It was connected by a chain with another rod in his room; the latter rod resting in a glass vessel, and being therefore insulated from the earth. On the 6th of August, 1753, a thunder cloud discharged itself against the external rod; the electricity passed downwards along the chain; on reaching the rod below, it was stopped by the glass vessel, darted to Richmann's head, which was about a foot distant, and killed him on the spot. Had a perfect communication existed between the lower rod and the earth, the lightning in this case would have expended itself harmlessly.

In 1749 Franklin proposed lightning conductors. He repeated his recommendation in 1753. He was opposed on two grounds. The

Abbé Nollet, and those who thought with him, considered it as impious to ward off heaven's lightnings, as for a child to ward off the chastening rod of its father. Others thought that the conductors would 'invite' the lightning to break upon them. A long discussion was also carried on as to whether the conductors should be blunt or pointed. Wilson advocated blunt conductors against Franklin, Cavendish, and Watson. He so influenced George III., hinting that the points were a republican device to injure his Majesty, that the pointed conductors on Buckingham House were changed for others ending in balls. Experience of the most varied kind has justified the employment of pointed conductors. In 1769 St. Paul's Cathedral was first protected.

The most decisive evidence in favour of conductors was obtained from ships; and such evidence was needed, to overcome the obstinate prejudice of seamen. Case after case occurred in which ships unprotected by conductors were singled out from protected ships, and shattered or destroyed by lightning. The conductors were at first made movable, being hoisted on the approach of a thunderstorm; but these were finally abandoned for the fixed lightning conductors devised by the late Sir Snow Harris. The saving of property and life by this obvious outgrowth of electrical research is incalculable.

§ 31. *The Returning Stroke.*

In the year 1779 Charles, Viscount Mahon, afterwards Earl Stanhope, published his 'Principles of Electricity.' On the title-page of the book stands the following remark:— 'This treatise comprehends an explanation of an electrical *returning stroke,* by which fatal effects may be produced even at a vast distance from the place where the lightning falls.'

Lord Mahon's experiments, which are models of scientific clearness and precision, will be readily understood by reference to the principles of electric induction, with which you are now so familiar.

It need only be noted here that whenever he speaks of a body being plunged in an 'electrical atmosphere,' he means that the body is exposed to the inductive action of a second electrified body, which latter he supposed to be surrounded by such an atmosphere.

A few extracts from his work will give a clear notion of the nature of his discovery:—

'I placed an insulated metallic cylinder, A B, fig. 55, within the electrical atmosphere of the prime conductor [P C] when charged,

Fig. 55.

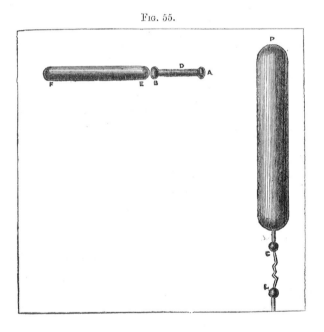

but beyond the striking distance. The distance between the near end A of the insulated metallic body and the side of the prime conductor was 20 inches. The body A B was of brass, of a cylindrical form, 18 inches long by 2 inches in diameter. I then placed another insulated brass body E F, 40 inches long by about $3\frac{3}{4}$ inches in diameter, with its end E at the distance of about one-tenth of an inch from the end B of the other metallic body A B. I electrified the prime conductor. All the time that it was receiving its *plus* charge of electricity there passed a great number of weak (red or purple) sparks from the end B of the near body A B into the end E of the remote body E F.'

Make clear to your mind the origin of this stream of weak red or purple sparks. It is obviously due to the inductive action of the prime conductor P C upon the body A B. The positive electricity of A B being repelled by the prime conductor, passed as a stream of sparks to E F.

'When the prime conductor, having received its full charge, came suddenly to discharge, with an explosion, its superabundant electricity on a large brass ball L, which was made to communicate with the earth, it always happened that the electrical fluid, which had been gradually expelled from the body A B and driven into the body E F, did suddenly return from the body E F into the body A B, in a strong and bright spark, at the very instant that the explosion took place upon the ball L.

'This I call the electrical *returning stroke.*'

For the two conductors Lord Mahon then substituted his own body and that of another person, both of them standing upon insulating stools. He continues thus:—

'I placed myself upon an insulating stool E (fig. 56), so as to have my right arm A at the distance of about 20 inches from a large prime conductor; another person, standing upon another insulating stool K, brought his right hand F within one-quarter of an inch of my left hand B.

'When the prime conductor began to receive its plus charge of electricity, we felt the electrical fluid running out of my hand B into his hand F.

'When we separated our hands B and F a little, the electricity passed between us in small sparks, which sparks increased in sharpness the farther we removed our hands B and F asunder, until we had brought them quite out of a striking distance. The intervals

Fig. 56.

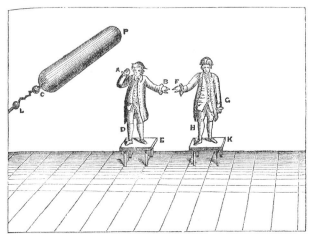

of time between these *departing sparks* increased also the more the distance between our hands B and F was increased, as must necessarily be the case.

'As soon as the prime conductor came suddenly to discharge its electricity upon the ball L, the superabundant electricity which the other person had received from my body did then return from him to me in a sharp spark, which issued from his hand F at the very instant that the explosion of the prime conductor took place upon the ball L.

'I still continued upon the insulating stool E, and I desired the other person to stand upon the floor. The returning stroke between us was *still stronger* than it had yet been. The reason of it was this:— the other person being no longer insulated, transmitted his superabundant electricity freely into the earth. I consequently became still more negative than before.

'Now, when the returning stroke came to take place, not only the electricity which had passed from my body into the body of the other person, but also the electricity which had passed from my body into the earth (through the other person), did suddenly return upon me from his hand F to my hand B, at the same instant that the discharge of the prime conductor took place upon the ball L. This caused the returning stroke to be stronger than before.'

Lord Mahon fused metals, and produced strong physiological effects by the return stroke.

In nature disastrous effects may be produced by the return stroke. The earth's surface, and animals or men upon it, may be powerfully influenced by one end of an electrified cloud. Discharge may occur at the other end, possibly miles away. The restoration of the electric equilibrium by the return shock may be so violent as to cause death.

This was clearly seen and illustrated by Lord Mahon. Fig. 57 is a reduced copy of his illustration. A B C is the electrified cloud, the two ends of which, A and C, come near the earth. The discharge occurs at C. A man at F is killed by the returning stroke, while the people at D, nearer to the place of discharge, but farther from the cloud, are uninjured.

With the view of still further testing your knowledge of induction, I have here copied a portion of this admirable essay; but the

FIG. 57.

entire memoir of Lord Mahon would constitute a most useful and interesting lesson in electricity.

For our own instruction we can illustrate the return shock thus:— Connect one arm of your universal discharger, fig. 49, with a conductor like c, fig. 20, and the other arm with the earth. Bring c within a few inches of your prime conductor, but not within striking distance; on working the machine a stream of feeble sparks will pass from point to point of the discharger. Let the prime conductor be discharged from time to time by an assistant; at every discharge the returning stroke is announced by a flash between the points of the discharger at s. If gun-cotton with a little fulminating powder scattered on it, or a fine silver wire, be introduced between the points of the discharger, the one is exploded and the other deflagrated.

The stream of repelled sparks first seen may be entirely abolished by establishing an *imperfect* connexion between the conductor c and the earth: a chain resting upon the dry table on which the conductor stands will do. The chain permits the feebler sparks to pass through it in preference to crossing the space s; but the returning stroke is too strong and sudden to find a sufficiently open channel through the table and chain, and on the discharge of the prime conductor the spark is seen.

It was the action of the return shock upon a dead frog's limbs, observed in the laboratory of Professor Galvani, that led to Galvani's experiments on animal electricity; and led further to the discovery, by Volta, of the electricity which bears his name.

§ 32. *The Leyden Battery, its Currents, and some of their Effects.*

In the ordinary Leyden battery described in § 19 all the inner coatings are connected together, and all the outer coatings are also connected together. Such a battery acts as a single large jar of extraordinary dimensions.

Wires are warmed by a moderate electric discharge; by augmenting the charge they are caused to glow; with a strengthened charge the metal is torn to pieces; fusion follows; and by still stronger charges the wires are reduced to metallic dust and vapour.

For such experiments the wire must be thin. Without resistance we can have no heat, and when the wire is thick we have little resistance. The mechanism of the discharge, as shown by the figures produced, is different in different wires. The figure produced by the dust of a deflagrated silver wire on white paper is shown in fig. 58.

Fig. 58.

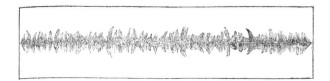

When the discharge of a powerful battery is sent through a long steel chain with the ends of its links unsoldered, the sparks between the unsoldered links carry the incandescent particles of the steel along with them. These are consumed in the air, a momentary blaze occurring along the entire chain. Chain cables have been fused by being made the channels of a flash of lightning.

Retaining our conception of an electric fluid, at this point we naturally add to it the conception of *a current*. It is the electric current which produces the effects just described. In many of our former experiments we had electricity at rest (static electricity), here we have electricity in motion (dynamic electricity).

Sending the current from a battery through a flat spiral (the primary) formed of fifty or sixty feet of copper wire, and placing within a little distance of it a second similar spiral (the secondary) with its ends connected; the passage of the current in the first spiral excites in the second a current, which is competent to deflagrate wires, and to produce all the other effects of the electrical discharge. Even when the spirals are some feet asunder, the shock produced by the secondary current is still manifest.

The current from the secondary spiral may be carried round a third; and this third spiral may be allowed to act upon a fourth, exactly as the primary did upon the secondary. A tertiary current is thus evoked by the secondary in the fourth spiral.

Carrying this tertiary current round a fifth spiral, and causing it to act inductively upon a sixth, we obtain in the latter a current of the fourth order. In this way we generate a long progeny of currents, all of them having the current sent from the battery through the first spiral, for a common progenitor. To Prof. Henry of the United States, and to Prof. Riess of Berlin, we are indebted for the investigation of the laws of these currents. These researches, however, were subsequent to, and were indeed suggested by, experiments of a similar character previously made by Faraday with Voltaic electricity.

Besides the electricity of friction and induction we have the following sources and forms of this power.

The contact of dissimilar metals produces electricity.

The contact of metals with liquids produces electricity.

A mere variation of the character of the contact of two bodies produces electricity.

Chemical action produces a continuous flow of electricity (Voltaic electricity).

Heat, suitably applied to dissimilar metals, produces a continuous flow of electricity (thermo-electricity).

The heating and cooling of certain crystals produce electricity (pyro-electricity).

The motion of magnets, and of bodies carrying electric currents, produces electricity (magneto-electricity).

The friction of sand against a metal plate produces electricity. The friction of condensed water-particles against a safety valve, or better still against a box-wood nozzle through which steam is driven, produces electricity (Armstrong's hydro-electric machine).

These are different manifestations of one and the same power; and they are all evoked by an equivalent expenditure of some other power.

Stars

Robert Stawell Ball*

We Try to Make a Map—The Stars are Suns—The Numbers of the
Stars—The Clusters of Stars—The Rank of the Earth as a Globe in
Space—The Distances of the Stars—The Brightness and Colour of
Stars—Double Stars—How we find what the Stars are Made of—The
Nebulae—What the Nebulae are Made of—Photographing the
Nebulae—Conclusion.

WE TRY TO MAKE A MAP.

THE group of bodies which cluster around our sun forms a little
island, so to speak, in the extent of infinite space. We may illustrate
this by a map in which we shall endeavour to show the stars placed
at their proper relative distances. We first open the compasses one
inch, and thus draw a little circle, which I intend to represent the
path followed by our earth, the sun being at the centre of the circle.
We are not going to put in all the planets. We take Neptune, the
outermost, at once. To draw its path I open the compasses to thirty
inches and draw a circle with that radius. That will do for our solar
system, though the comets no doubt will roam beyond these limits.

---

* The original version appeared in: Robert Stawell Ball, Lecture 6, *Star-Land. Being Talks with Young People About the Wonders of the Heavens*, London, 1889, pp. 297–356.

To complete our map we ought of course to put in some stars. There are a hundred million to choose from, and we shall begin with the brightest. It is often called the Dog star, but astronomers know it better as Sirius. Let us see where it is to be placed on our map. Sirius is beyond Neptune, so it must be outside somewhere. Indeed, it is a good deal further off than Neptune; so I try at the edge of the drawing-board: I have got a method of making a little calculation that I do not intend to trouble you with, but I can assure you that the results it leads me to are quite correct; they show me that this board is not big enough. We must ask the Royal Institution to provide a larger board in this room. But could a board which was big enough fit into this room? Here, again, I make my little calculations, and I find that the room would not hold a board sufficiently great; in fact, if I put the sun here at one end, with its planets around it, Sirius would be too near if it were at the opposite corner of the room. The board would have to go out through the wall of the theatre, out through London. Indeed, big as London is, it would not be large enough to contain the drawing-board that I should require. it [sic] would have to stretch about twenty miles from where we are now assembled. We may therefore dismiss any hope of making a practicable map of our system on this scale if Sirius is to have its proper place. Let us, then, take some other star. We shall naturally try with the nearest of all. It is one that we do not know in this part of the world, but those who live in the southern hemisphere are well acquainted with it. The name of this star is Alpha Centauri. Even for this star, we should require a drawing three or four miles long if the distance from the earth to the sun is to be taken as one inch. You see what an isolated position our sun and his planets occupy. The members of the family are all close together, and the nearest neighbours are situated at enormous distances. There is a good reason for this separation. The stars are very pretty where they lie, but, as they might be very troublesome neighbours if they were close to our system, it is well they are so far off; they would be constantly making disturbance in the sun's family if they were near at hand. Sometimes they would be dragging us into hot water by bringing us too close to the sun, or producing a coolness by pulling us away from the sun, which would be quite as disagreeable.

THE STARS ARE SUNS.

We are about to discuss one of the grandest truths in the whole of nature. We have had occasion to see that this sun of ours is a magnificent globe immensely larger than the greatest of his planets, while the greatest of these planets is immensely larger than this earth; but now we are to learn that our sun is, after all, only a star not nearly so bright as many of those which shine over our heads every night. We are comparatively close to the sun, so that we are able to enjoy his beautiful light and cheering heat. All those other myriads of stars are each of them suns, and the splendour of those distant suns is often far greater than that of our own. We are, however; so enormously far from them that they have dwindled down to insignificance. To judge impartially between our star or sun and such a sun or star as Sirius we should stand half-way between the two: it is impossible to make a fair estimate when we find ourselves situated close up to one star and a million times as far from the other. When we make allowance for the imperfections of our point of view, we are enabled to realise the majestic truth that our sun is no more than a star, and that the other stars are no less than suns. This gives us an imposing idea of the extent and the magnificence of the universe in which we are situated. Look up to the sky at night—you will see a host of stars: try to think that every one of them is itself a sun. It may probably be that those suns have planets circulating around them, but it is hopeless for us to expect to see such planets. Were you standing on one of those stars and looking towards our system, you would not perceive the sun to be the brilliant and gorgeous object that we know so well. If you could see him at all, he would merely seem like a star, not nearly so bright as many of those you can see at night. Even if you had the biggest of telescopes to aid your vision, you could never discern from one of these bodies the planets which surround the sun. No astronomer in the stars could see Jupiter, even if his sight were a thousand times as good or his telescopes a thousand times as powerful as any sight or telescope that we know. So minute an object as our earth would, of course, be still more hopelessly beyond the possibility of vision.

THE NUMBERS OF THE STARS.

To count the stars involves a task which lies beyond the power of man to fully accomplish. Even without the aid of any telescope, we can see a great multitude of stars from this part of the world. There are also many constellations in the southern hemisphere which never appear above our horizon. If, however, we were to go to the equator, then, by waiting there for a twelvemonth, all the stars in the heavens would have been successively exposed to view. An astronomer, Houzeau, who had the patience to count them, enumerated about 6,000. This is the naked-eye estimate of the star population of the heavens; but if, instead of relying on unassisted eyes, you get the assistance of a little telescope, you will be astounded at the enormous multitude of stars which are disclosed.

An ordinary opera-glass is a very useful instrument for looking at the stars in the heavens, as well as at the stars of another description, to which it is more commonly applied. You will be amazed to find that the heavens teem with additional multitudes of stars that the opera-glass will reveal. Any part of the sky may be observed; but, just to give an illustration, I shall take one special region, namely, that of the Great Bear (Fig. 80). The seven well-known stars

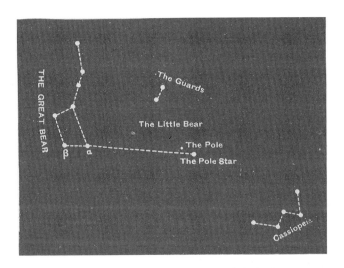

Fig. 80.—The Great Bear and the Pole.

are here shown, four of which form a sort of oblong, while the other three represent the tail. In an Appendix I shall describe the services of the Great Bear as a guide to a knowledge of the constellations. I would like you to make this little experiment. On a fine clear night, count how many stars there are within this oblong; they are all very faint, but you will be able to see a few, and, with good sight, and on a good night, you may see perhaps ten. Next take your opera-glass and sweep it over the same region: if you will carefully count the stars it shows you will find fully 200; so that the opera-glass has, in this part of the sky, revealed nearly twenty times as many stars as could be seen without its aid. As 6,000 stars can be seen by the eye, all over the heavens, we may fairly expect that twenty times that number—that is to say, 120,000 stars—could be shown by the opera-glass over the entire sky. Let us go a step further, and employ a telescope, the object-glass of which is three inches across. This is a useful telescope to have, and, if a good one, will show multitudes of pleasing objects, though an astronomer would not consider it very powerful. An instrument like this, small enough to be carried in the hand, has been applied to the task of enumerating the stars in the northern half of the sky, and 320,000 stars were counted. Indeed, the actual number that might have been seen with it is considerably greater, for when the astronomer Argelander made this memorable investigation he was unable to reckon many of the stars in localities where they lay very close together. This grand count only extended to half the sky, and, assuming that the other half is as populous, we see that a little telescope like that we have supposed will show over the sky a number of stars which exceeds that of the population of any city in England except London. It exhibits more than one hundred times as many stars as our eyes could possibly reveal. Still, we are only at the beginning of the count: the really great telescopes add largely to the number. There are multitudes of stars which, in small telescopes, we cannot see, but which are distinctly visible from our great observatories. That telescope would be still but a comparatively small one which would show as many stars in the sky as there are people living in this mighty city of London; and with the greatest instruments, the tale of stars has risen to a number far greater than that of the entire population of Great Britain.

In addition to those stars the largest telescopes show us, there are myriads which make their presence evident in a wholly different way. It is only in quite recent times that an attempt has been made to develop fully the powers of photography in representing the celestial objects. On a photographic plate which has been exposed to the sky in a great telescope the stars are recorded in their thousands. Many of these may, of course, be observed with a good telescope, but there are not a few others which no one ever saw in a telescope, which apparently no one ever could see, though the photograph is able to show them. We do not, however, employ a camera like that which the photographer uses who is going to take your portrait. The astronomer's plate is put into his telescope, and then the telescope is turned towards the sky. On that plate the stars produce their images, each with its own light. Some of these images are excessively faint, but we give a very long exposure of an hour or two hours; sometimes so much as four hours' exposure is given to a plate so sensitive that a mere fraction of a second would sufficiently expose it during the ordinary practice of taking a photograph in daylight. We thus afford sufficient time to enable the fainter objects to indicate their presence upon the sensitive film. Even with an exposure of a single hour a picture exhibiting 16,000 stars has been taken by Mr. Isaac Roberts, of Liverpool. Yet the portion of the sky which it represents is only one ten-thousandth part of the entire heavens. It should be added that the region which Mr. Roberts has photographed is furnished with stars in rather exceptional profusion.

Here, at last, we have obtained some conception of the sublime scale on which the stellar universe is constructed. Yet even these plates cannot represent all the stars that the heavens contain. We have every reason for knowing that with larger telescopes, with more sensitive plates, with more prolonged exposures, ever fresh myriads of stars will be brought within our view.

You must remember that every one of these stars is truly a sun, a lamp as it were, which doubtless gives light to other objects in its neighbourhood as our sun sheds light upon this earth and the other planets. In fact, to realise the glories of the heavens you should try to think that the brilliant points you see are merely the luminous points of the otherwise invisible universe.

Standing one fine night on the deck of a Cunarder we passed in open ocean another great Atlantic steamer. The vessel was near enough for us to see not only the light from the mast-head but also the little beams from the several cabin ports; but we could see nothing of the ship herself.

Her very existence was only known to us by the twinkle of these lights. Doubtless her passengers could see, and did see, the similar lights from our own vessel, and they doubtless drew the correct inference that these lights indicated a great ship.

Consider the multiplicity of beings and objects in a ship: the captain and the crew, the passengers, the cabins, the engines, the boats, the rigging, and the stores. Think of all the varied interests there collected and then reflect that out on the ocean, at night, the sole indication of the existence of this elaborate structure was given by the few beams of light that happened to radiate from it. Now raise your eyes to the stars, there are the twinkling lights. We cannot see what those lights illuminate, nor can we conjecture what untold wealth of non-luminous bodies may also lie in their vicinity; we may, however, feel certain that just as the few gleaming lights from a ship are utterly inadequate to give a notion of the nature and the contents of an Atlantic steamer, so are the twinkling stars utterly inadequate to give even the faintest conception of the extent and the interest of the universe. We merely see self-luminous bodies, but of the multitudes of objects and the elaborate systems of which these bodies are only the conspicuous points we see nothing and we know nothing. We are, however, entitled to infer from an examination of our own star—the sun—and of the beautiful system by which it is surrounded, that these other suns may be also splendidly attended. This is quite as reasonable a supposition as that a set of lights seen at night on the Atlantic Ocean indicate the existence of a fine ship.

THE CLUSTERS OF STARS.

On a clear night you can often see, stretching across the sky, a track of faint light, which is known to astronomers as the "Milky Way." It extends below the horizon and then round the earth to form a girdle about the heavens. When we examine the Milky

Way with a telescope we find, to our amazement, that it consists of myriads of stars, so small and so faint that we are not able to distinguish them individually, we merely see the glow produced from their collective rays. Remembering that our sun is a star, and that the Milky Way surrounds us, it would almost seem as if our sun were but one of the host of stars which form this cluster.

There are also other clusters of stars, some of which are most exquisitely beautiful telescopic spectacles. I may mention a celebrated pair of these objects which lie in the constellation of Perseus. The sight of these in a great telescope is so imposing that no one who is fit to look through a telescope could resist a shout of wonder and admiration when first they burst on his view. But there are other clusters. Here is a picture of one which is known as the "Globular Cluster in the Centaur" (Fig. 81). It consists of a ball of stars, so far off, that however large these several suns may actually be, they have dwindled down to extremely small points of light. A homely illustration may serve to show the appearance which a globular cluster presents in a good telescope. I take a pepper-castor

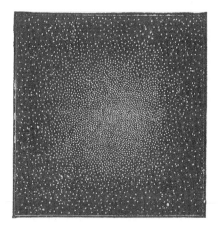

Fig. 81.—Globular Cluster in the Centaur.

and on a sheet of white paper I begin to shake out the pepper until there is a little heap at the centre and other grains are scattered loosely about. Imagine that every one of those grains of pepper was to be transformed into a tiny electric light, and then you have some idea of what a cluster of stars would look like when viewed through a telescope of sufficient power. There are multitudes of such clusters scattered through the depths of space. They require our biggest telescopes to show them adequately. We have seen that our sun is a star, being only one of a magnificent cluster that form the Milky Way. We have also seen that there are other clusters scattered through the length and depth of space. It is thus we obtain a notion of the rank which our earth holds in the scheme of things celestial.

THE RANK OF THE EARTH AS A GLOBE IN SPACE.

Let me give an illustration with the view of explaining more fully the nature of the relation which the earth bears to the other globes which abound through space, and you must allow me to draw a little upon my imagination. I shall suppose that Her Majesty's mails extend not only over this globe, but that they also communicate with other worlds; that postal arrangements exist between Mars and the earth, between the sun and Orion—in fact, everywhere throughout the whole extent of the universe. We shall consider how our letters are to be addressed. Let us take the case of Mr. John Smith, merchant, who lives at 1,001, Piccadilly; and let us suppose that Mr. John Smith's business transactions are of such an extensive nature that they reach not only all over this globe, but away throughout space. I shall suppose that the firm has a correspondent residing—let us say in the constellation of the Great Bear; and when this man of business wants to write to Mr. Smith from these remote regions, what address must he put upon the letter, so that the Postmaster-General of the universe shall make no mistake about its delivery? He will write as follows:—

MR. JOHN SMITH,
1,001, Piccadilly,
London,
England,
Europe,
Earth,
Near the Sun,
Milky Way,
The Universe.

Let us now see what the several lines of this address mean. Of course we put down the name of Mr. John Smith in the first line, and then we will add "1 001, Piccadilly," for the second; but as the people in the Great Bear are not likely to know where Piccadilly is, we shall add "London" underneath. As even London itself cannot be well known everywhere, it is better to write "England" underneath. This would surely find Mr. John Smith from any post-office on this globe. From other globes, however, the supreme importance of England may not be so immediately recognised, and therefore it is as well to add another line "Europe." This ought to be sufficient, I think, for any post-office in the solar system. Europe is big enough to be visible from Mars or Venus, and should be known to the post-office people there, just as we know and have names for the continents on Mars. But further away there might be a little difficulty: from Uranus and Neptune the different regions on our earth can never have been distinguished, and therefore we must add another line to indicate the particular globe of the solar system which contains Europe. Mark Twain tells us that there was always one thing in astronomy which specially puzzled him, and that was to know how we found out the names of the stars. We are, of course, in hopeless ignorance of the name by which this earth is called among other intelligent beings elsewhere who can see it. I can only adopt the title of "Earth," and therefore I add this line. Now our address is so complete that from anywhere in the solar system—from Mercury, from Jupiter, or Neptune—there ought to be no mistake about the letter finding its way to Mr. John Smith. But from his correspondent in the Great Bear this address would be still incomplete; they cannot see our earth from thence, and even the sun himself only looks like a small star—like one, in fact, of thousands of stars elsewhere. However, each star can be distinguished, and our sun may, for instance, be recognised from the Great Bear by some designation. We shall add the line "Near the Sun," and then I think that from this constellation, or from any of the other stars around us, the address of Mr. John Smith may be regarded as complete. But Mr. Smith's correspondence may be still wider. He may have an agent living in the cluster of Perseus or on some other objects still fainter and more distant; then "Near the Sun" is utterly inadequate as a concluding line to the address, for the sun, if it can be seen at all from thence, will be only of the significance of an excessively minute star, no more to be designated by a special name than are the several leaves on the trees of a forest. What this distant correspondent will be acquainted with is not the earth or the sun, but only the cluster of stars among which the sun is but a unit. Again we use our own name to denote the cluster, and we call it the "Milky Way." When we add this line, we have made the address of Mr. John Smith as complete as circumstances permit it to be. I think a letter posted to him anywhere ought to reach its destination. For completeness, however, we will finish up with one line more—"*The Universe.*"

THE DISTANCES OF THE STARS.

I must now tell you something about the distances of the stars. I shall not make the attempt to explain fully how astronomers make such measurements, but I will give you some notion of how it is done. You may remember I showed you how we found the distance of a globe that was hung from the ceiling. The principle of the method for finding the distance of the star is somewhat similar, except that we make the two observations not from the two ends of a table, not even from the two sides of the earth, but from two opposite points on the earth's orbit, which are therefore at a distance of 186,000,000 miles. Imagine that on Midsummer Day, when standing on the earth here, I measure with a piece of card the angle between the star and the sun. Six months later, on Midwinter Day, when the earth is at the opposite point of its orbit, I again measure the angle between the same star and the sun, and we can now determine the star's distance by making a triangle. I draw a line a foot long, and we will take this foot to represent 186,000,000 miles, the distance between the two stations; then, placing the cards at the corners, I rule the two sides and complete the triangle, and the star must be at the remaining corner; then I measure the sides of the triangle and find how many feet they contain, and recollecting that each foot corresponds to 186,000,000 miles, we discover the distance of the star. If the stars were comparatively near us, the process would be a very simple one; but, unfortunately, the stars are so extremely far off, that this triangle, even with a base of only one foot, must have its sides many miles long. Indeed, astronomers will tell you that there is no more delicate or troublesome work in the whole of their science than that of discovering the distance of a star.

In all such measurements we take the distance from the earth to the sun as a conveniently long measuring rod, whereby to express the results. The nearest stars are still hundreds of thousands of times as far off as the sun. Let us ponder for a little on magnitudes so vast. We shall first express them in miles. Taking the sun's distance to be 93,000,000 miles, then the distance of the nearest fixed star is about twenty millions of millions of miles—that is to say, we express this distance by putting down a 2 first, and then writing

thirteen cyphers after it. It is, no doubt, easy to speak of such figures, but it is a very different matter when we endeavour to imagine the awful magnitude which such a number indicates. I must try
to give some illustrations which will enable you to form a notion of
it. At first I was going to ask you to try and count this number, but
when I found it would require at least 300,000 years, counting day
and night without stopping before the task was over, it became
necessary to adopt some other method.

When lately in Lancashire I was kindly permitted to visit a cotton mill, and I learned that the cotton yarn there produced in a
single day would be long enough to wind round this earth twenty-
seven times at the equator. It appears that the total production of
cotton yarn each day in all the mills together would be on the average about 155,000,000 miles. In fact, if they would only spin about
one-fifth more, we could assert that Great Britain produced enough
cotton yarn every day to stretch from the earth to the sun and back
again! It is not hard to find from these figures how long it would
take for all the mills in Lancashire to produce a piece of yarn long
enough to reach from our earth to the nearest of the stars. If the
spinners worked as hard as ever they could for a year, and if all
the pieces were then tied together, they would extend to only a
small fraction of the distance; nor if they worked for ten years, or
for twenty years, would the task be fully accomplished. Indeed,
upwards of 400 years would be necessary before enough cotton
could be grown in America and spun in this country to stretch over
a distance so enormous. All the spinning that has ever yet been
done in the world has not formed a long enough thread!

There is another way in which we can form some notion of the
immensity of these sidereal distances. You will recollect that, when
we were speaking of Jupiter's moons (p. 205) [not included here], I
told you of the beautiful discovery which their eclipses enabled
astronomers to make. It was thus found that light travels at the
enormous speed of about 185,000 miles per second. It moves so

quickly that within a single second a ray would flash two hundred times from London to Edinburgh and back again.

We said that a meteor travels one hundred times as swiftly as a rifle-bullet; but even this great speed seems almost nothing when compared with the speed of light, which is 10,000 times as great. Suppose some brilliant outbreak of light were to take place in a distant star—an outbreak which would be of such intensity that the flash from it would extend far and wide throughout the universe. The light would start forth on its voyage with terrific speed. Any neighbouring star which was at a distance of less than 185,000 miles would, of course, see the flash within a second after it had been produced. More distant bodies would receive the intimation after intervals of time proportional to their distances. Thus, if a body were 1,000,000 miles away the light would reach it in five or six seconds, while over a distance as great as that which separates the earth from the sun the news would be carried in about eight minutes. We can calculate how long a time must elapse ere the light shall travel over a distance so great as that between the star and our earth. You will find that from the nearest of the stars the time required for the journey will be over three years. Ponder on all that this involves. That outbreak in the star might be great enough to be visible here, but we could never become aware of it till three years after it had happened. When we are looking at such a star to-night we do not see it as it is at present, for the light that is at this moment entering our eyes has travelled so far that it has been three years on the way, therefore, when we look at the star now we see it as it was three years previously. In fact, if the star was to go out altogether, we might still continue to see it twinkling away for a period of three years longer, because a certain amount of light was on its way to us at the moment of extinction, and so long as that light keeps arriving here, so long shall we see the star showing as brightly as ever. When, therefore, you look at the thousands of stars in the sky to-night, there is not one that you see as it is now, but as it was years ago.

I have been speaking of the stars that are nearest to us, but there are others much farther off. It is true we cannot find the distance of these more remote objects with any degree of accuracy, but we can convince ourselves how great that distance is by the following reasoning. Look at one of the brightest stars. Try to conceive that the object was drawn away further into the depths of space, until it was ten times as far from us as it is at present, it would still remain bright enough to be recognised in quite a small telescope; even if it were taken to one hundred times its original distance it would not have withdrawn from the view of a good telescope; while if it retreated one thousand times as far as it was at first it would still be a recognisable point in our mightiest instruments. Among the stars which we can see in our telescopes, we feel confident there must be many from which the light has taken hundreds of years, or even thousands of years, to arrive here. When, therefore, we look at such objects, we see them, not as they are now, but as they were ages ago; in fact, a star might have ceased to exist for thousands of years, and still be seen by us every night as a twinkling point in our great telescopes.

Remembering these facts, you will, I think, look at the heavens with a new interest. There is a bright star, Vega or Alpha Lyrae, a beautiful gem, but so far off that the light from it which we now see started before many of my audience were born. Suppose that there are astronomers residing on worlds amid the stars, and that they have sufficiently powerful telescopes to view this globe, what do you think they will observe? They will not see our earth as it is at present, they will see us as we were years ago. There are stars from which, if England could now be seen, the whole of the country would be observed at this present moment to be in a great state of excitement at a very auspicious event. Distant astronomers might observe a great procession in London amid the enthusiasm of a nation, and they could watch the coronation of a youthful queen. There are other stars still further off, from which, if the inhabitants had good enough telescopes, they would now see a mighty battle in progress not far from Brussels: they would see one army dashing itself time after time against the immovable ranks of the other. I do not think they would be able to hear the ever-memorable, "Up, Guards, and at them!" but there can be no doubt that there are stars so far away that the rays of light which started from the earth on the day of the Battle of Waterloo are only just arriving there. Further off still, there are stars from which a bird's-eye view could be taken at this very moment of the signing of Magna Charta. There are even stars from which England, if it could be seen at all, would now appear, not as the great England we know, but as a country covered by dense forests, and inhabited by painted savages, who waged incessant war with wild beasts that roamed through the island. The geological problems that now puzzle us would be quickly solved could we only go far enough into space and had we only powerful enough telescopes. We should then be able to view our earth through the successive epochs of past geological time: we should be actually able to see those great animals whose fossil remains are treasured in our museums, tramping about over the earth's surface, splashing across its swamps, or swimming with broad flippers through its oceans. Indeed, if we could view our own earth reflected from mirrors in the stars, we could still see Moses crossing the Red Sea, or Adam and Eve being expelled from Eden.

So important is the subject of star distance that I am tempted to give one more illustration in order to bring before you some conception of how vast that distance truly is. I shall take, as before, the nearest of the stars so far as known to us, and I hope to be forgiven for taking an illustration of a practical and a commercial kind instead of one more purely scientific. I shall suppose that a railway is about to be made from London to Alpha Centauri. The length of that railway, of course, we have already stated: it is twenty billions of miles. So I am now going to ask your attention to the simple question as to the fare which it would be reasonable to charge for the journey. We shall choose a very cheap scale on which to compute the fare. The parliamentary rate here is, I believe, a penny for every mile. We will make our interstellar railway fares much less even than this; we shall arrange to travel at the rate of one hundred miles for every penny. That, surely, is moderate enough. If our fares were so low that the journey from London to Edinburgh only cost four-pence, then even the most unreasonable passenger would be surely contented. On these terms how much do you think the fare from London to this star ought to be? I know of one way in which to make the answer intelligible. There is a National Debt with which your fathers are, unhappily, only too well acquainted: you will know quite enough about it yourselves in those days when you have to pay income tax. This Debt is so vast that the interest upon it alone is about sixty thousand pounds a day, the whole amount of the National Debt being seven hundred and thirty-six millions of pounds (April, 1887).

If you went to the booking-office with the whole of this mighty sum in your pocket—but stop a moment: could you carry it in your pocket? Certainly not, if it were in sovereigns. You would find that after your pocket had as many sovereigns as it could conveniently hold there would still be some left—so many, indeed, that it would be necessary to get a cart to help you on with the rest. When the cart had as great a load of sovereigns as the horse could draw there would be still some more, and you would have to get another cart; but ten carts, twenty carts, fifty carts, would not be enough. You would want five thousand carts before you would be able to move off towards the station with your money. When you did get there and asked for a ticket at the rate of 100 miles for a penny, do you think you would get any change back? No doubt some little time would be required to count the money, but when it was counted the clerk would tell you that it was not enough, that he must have nearly a hundred millions of pounds more.

That will give some notion of the distance of the nearest star, and we may multiply it by ten, by one hundred, and even by one thousand, and still not attain to the distance of some of the more distant stars that the telescope shows us.

On account of the immense distances of the stars we can only perceive them to be mere points of light. You will never see a star to be a globe with marks on it like the moon, or like one of the planets—in fact, the better the telescope the smaller does a star seem, though, of course, its brightness is increased with every increase in the light-grasping power of the instrument.

THE BRIGHTNESS AND COLOUR OF STARS.

Another point to be noticed is the arrangement of stars in classes, according to their lustre. The brightest stars, of which there are about twenty, are said to be of the first magnitude. The stars just inferior to the first magnitude are ranked as the second; and those just inferior to the second are estimated as the third; and so on. The smallest stars that your unaided eyes will show you are of about the sixth magnitude. Then the telescope will reveal stars still fainter and fainter, down to what we term the seventeenth or eighteenth magnitudes, or even lower still. The number of stars of each magnitude increases very much in the classes of small stars.

There is one of the larger stars which should be specially mentioned. It is Sirius, or the Dog-star, which is much brighter than any other in the heavens. Most of the stars are white, but many are of a somewhat ruddy hue. There are a few telescopic stars which are intensely red, some exhibit beautiful golden tints, while others are blue or green.

There are some curious stars which regularly change their brilliancy. Let me try to illustrate the nature of these variables. Suppose that you were looking at a street gas-lamp from a very long distance, so that it seemed a little twinkling light; and suppose that someone was attending to turn the cock up and down. Or, better still, imagine a little machine which would act regularly so as to keep the light first of all at its full brightness for two days and a half, and then gradually turn it down until in three or four hours it declines to a feeble glimmer. In this low state the light remains for twenty minutes; then during three or four hours the gas is to be slowly and gradually turned on again until it is full. In this condition the light will remain for two days and a half, and then the same series of changes is to commence. This would be a very odd form of gas-lamp. There would be periods of two days and a half during which it would remain at its full; these would be separated by intervals of about seven hours, when the slow turning down and turning up again would be in progress.

The imaginary gas-lamp is exactly paralleled by a star Algol, in the constellation of Perseus (Fig. 82), which goes through the series of changes I have indicated. Ordinarily speaking, it is a bright star of the second magnitude, and whatever be the cause, the star performs its variations with marvellous uniformity. In fact, Algol has always arrested the attention of those who observed the heavens, and in early times was looked on as the eye of a Demon. There are many other stars which also change their brilliancy. Most of them require much longer periods than Algol, and sometimes a new star which nobody has ever seen before will suddenly kindle into brilliancy.

DOUBLE STARS.

Whenever you have a chance of looking at the heavens through a telescope, you should ask to be shown what is called *a double star*. There are many stars in the heavens which present no remarkable appearance to the unaided eye, but which a good telescope at once

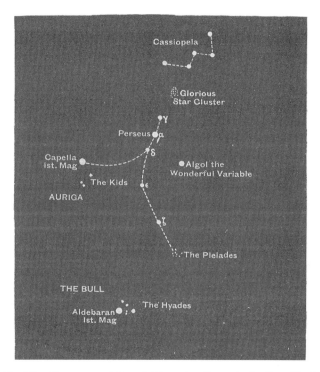

Fig. 82.—Perseus and its neighbouring Stars, including Algol.

shows to be of quite a complex nature. These are what we call double stars, in which the two little points of light are placed so close together that the unaided eye is unable to separate them. Under the magnifying power of the telescope, however, they are seen to be distinct. In order to give some notion of what these objects are like, I shall briefly describe three of them. The first lies in that best known of constellations, the Great Bear. If you look at his tail, which consists of three stars, you will see that near the middle one of the three a small star is situated; we call this star Alcor, but it is the brighter one near Alcor to which I specially call your attention. The sharpest eye would never suspect that this object was composed of two stars placed close together. Even a small telescope will, however, show this to be the case, and this is the easiest and the first observation that a young astronomer should make when beginning to turn a telescope to the heavens. Of course you will not imagine that I mean Alcor to be the second component of the double star; it is the bright star near Alcor which is the double. Here are two marbles, and these marbles are fastened an inch apart. You can see them, of course, to be separate; but if the pair were moved further and further away, then you would soon not be able to distinguish between them, though the actual distance between the marbles had not altered. Look at these two wax tapers which are now lighted; the little flames are an inch apart. You would have to view them from a station a third of a mile away if the distance between the two flames were to appear the same as that between the two components of this double star. Your eye would never be able to discriminate between two lights only an inch apart at so great a distance; a telescope would, however, enable you to do so, and this is the reason why we have to use telescopes to show us double stars.

You might look at that double star year after year throughout the course of a long life without finding any appreciable change in the relative positions of its components. But we know that there is no such thing as rest in the universe; even if you could balance a body so as to leave it for a moment at rest, it would not stay there, for the simple reason that all the bodies around it in every direction are pulling at it, and it is certain that the pull in one direction will preponderate so that move it must. Especially is this true in the case of two suns like those forming a double star. Placed comparatively near each other they could not remain permanently in that position; they must gradually draw together and come into collision with an awful crash. There is only one way by which such a disaster could be obviated. That is by making one of these stars revolve around the other just as the earth revolves around the sun, or the moon revolves around the earth. There must, therefore, be some motion going on in every genuine double star, whether we have been able to see that motion or not.

Let us now look at another double star of a different kind. This time it is in the constellation of Gemini. The heavenly twins are called Castor and Pollux. Of these, Castor is a very beautiful double star, consisting of two bright points, a great deal closer together than were those in the Great Bear; consequently a better telescope is required for the purpose of showing them separately. Castor has been watched for many years, and it can be seen that one of these stars is slowly revolving around the other; but it takes a very long time, amounting to hundreds of years, for a complete circuit to be accomplished. This seems very astonishing, but when you remember how exceedingly far Castor is from us, you will see that that pair of stars which seems so close together that it requires a telescope to show that they are distinct must indeed be separated by hundreds of millions of miles. Let us try to conceive our own system transformed into a double star. If we took our outermost planet—Neptune—and enlarged him a good deal, and then heated him sufficiently to make him glow like a sun, he would still continue to revolve round our sun at the same distance, and thus a double star would be produced. An inhabitant of Castor who turned his telescope towards us, would be able to see the sun as a star. He would not, of course, be able to see the earth, but he might see Neptune like another small star close up to the sun. If generations of astronomers in Castor continued their observations of our system, they would find a binary star, of which one component took a century and a half to go round the other. Need we then be surprised that when we look at Castor we observe movements equally deliberate?

There is often so much glare from the bright stars seen in a telescope, and so much twinkling in some states of the atmosphere, that stars appear to dance about in rather a puzzling fashion, especially to one who is not accustomed to astronomical observations. I remember hearing how a gentleman once came to visit an observatory. The astronomer showed him Castor through a powerful telescope as a fine specimen of a double star, and then, by way of improving his little lesson, the astronomer mentioned that one of these stars was revolving around the other. "Oh, yes," said the visitor, "I saw them going round and round in the telescope." He would, however, have had to wait for a few centuries with his eye to the instrument before he would have been entitled to make this assertion.

Double stars also frequently delight us by giving beautifully contrasted colours. I dare say you have often noticed the red and the green lights that are used on railways in the signal lamps. Imagine one of those red and one of those green lights away far up in the sky and placed close together, then you would have some idea of the appearance that a coloured double star presents, though, perhaps, I should add that the hues in the heavenly bodies are not so vividly contrasted as are those which our railway people find necessary. There is a particularly beautiful double star of this kind in the constellation of the Swan. You could make an imitation of it by boring two holes, with a red-hot needle, in a piece of card, and then covering one of these holes with a small bit of the topaz-coloured gelatine with which Christmas crackers are made. The other star is to be similarly covered with blue gelatine. A slide made on this principle placed in the lantern gives a very good representation of these two stars on the screen. There are many other coloured doubles besides this one; and, indeed, it is noteworthy that we hardly ever find a blue or a green star by itself in the sky: it is always as a member of one of these pairs.

HOW WE FIND WHAT THE STARS ARE MADE OF.

Here is a piece of stone. If I wanted to know what this stone was composed of, I should ask a chemist to tell me. He would take it into his laboratory, and first crush it up into powder, and then, with his test tubes, and with the liquids which his bottles containt and his weighing scales, and other apparatus, he will, tell all about it: there is so much of this, and so much of that, and plenty of this, and none at all of that. But now, suppose you ask this chemist to tell you what the sun is made of, or one of the stars. Of course, you have not a sample of it to give him; how then, can he possibly find out anything about it? Well, he can find out something, and this is the wonderful discovery that I want to explain to you. We now put down the gas, and I kindle a brilliant red light. Perhaps some of those whom I see before me have occasionally ventured on the somewhat dangerous practice of making fireworks. If there is any boy here who has ever made sky-rockets, and put the little balls into the top, which are to burn with such vivid colours when the explosion takes place, he will know that the substance which tinged that red fire must have been *strontium*. He will recognise it by the colour; because *strontium* gives that red light which nothing else will give. Here are some of these lightning papers, as they are called; they are very pretty and very harmless; and these, too, give brilliant red flashes as I throw them. The red tint has, no doubt, been given by *strontium* also. You see we recognised the substance simply by the colour of the light it produced when burning.

Perhaps some of you have tried to make a ghost at Christmas time by dressing up in a sheet, and bearing in your hand a ladle blazing with a mixture of common salt and spirits of wine, the effect produced being a most ghostly one. Some mammas will hardly thank me for this suggestion, unless I add that the ghost must walk about cautiously, for otherwise the blazing spirit would be very apt to produce conflagrations of a kind quite different from those intended. However, by the kindness of Professor Dewar, I am enabled to show the phenomenon on a splendid scale, and also free from all danger. I kindle a vivid flame of an intensely yellow colour, which I think the ladies will unanimously agree is not at all becoming to their complexions, while the pretty dresses have lost all their variety of colours. Here is a nice bouquet, and yet you can hardly distinguish the green of the leaves from the brilliant colours of the flowers, except by trifling difference of shade. Expose to this light a number of pieces of variously coloured ribbon, pink and red and green and blue, and all their beauty is gone; and yet we are told that this yellow is a perfectly pure colour; in fact the purest colour that can be produced. I think we have to be thankful that the light which our good sun sends us does not possess purity of that description. There is just one substance which will produce that yellow light; it is a curious metal called sodium—a metal so soft that you can cut it with a knife, and so light that it will float on water; while, still more strange, it actually takes fire the moment it is dropped on the water. It is only in a chemical laboratory that you will be likely to meet with the actual metallic sodium, yet in other forms the substance is one of the most abundant in nature. Indeed, common salt is nothing but sodium closely united with a most poisonous gas, a few respirations of which would kill you. But you see this strange metal and this noxious gas, when united, become simply the salt for our eggs at breakfast. This simple yellow light, wherever it is seen, either in the flame of spirits of wine mixed with salt or in that great blaze at which we have been looking, is a characteristic of sodium. Wherever you see that particular kind of light, you know that sodium must have been present in the body from which it came. We have accordingly learned to recognise two substances, namely, *strontium* and *sodium,* merely by the light which they give out when heated so as to burn. To these we may add a third. Here is a strip of white metallic ribbon. It is called magnesium. It seems like a bit of tin at the first glance, but, indeed, it is a very different thing from tin; for, look, when I hold it in the spirit-lamp, the strip of metal immediately takes fire, and burns with a white light so dazzling that it pales the gas-flames to insignificance. There is no other substance that will, when kindled, give that particular kind of light which we see from magnesium. I can recommend this little experiment as quite suitable for trying at home; you can buy a bit of magnesium ribbon for a trifle at the optician's; it cannot explode or do any harm, nor will you get into any trouble with the authorities provided you hold it when burning over a tray or a newspaper, so as to prevent the white ashes from falling on the carpet.

There are, in nature, a number of simple substances called elements. Every one of these, when ignited under suitable conditions, emits a light which belongs to it alone and by which it can be distinguished from every other substance. I do not say that we can try the experiments in the simple way I have here indicated. Many of the substances will only yield their light so as to be recognisable by much more elaborate artifices than those which have sufficed for us. But you see that the method affords a means of finding out the actual substances present in the sun or in the stars. One practical difficulty is, that each of the heavenly bodies contains a number of different elements; so that in the light it sends us the hues arising from distinct substances are all blended into one beam. The first thing to be done is to get some way of splitting up a beam of light, so as to discover the components of which it is made. You might have a skein of silks of different hues all tangled together, and this would be like the sunbeam as we receive it in its unsorted condition. How shall we untangle the light from the sun or a star? I will show you by a simple experiment. Here is a beam from the electric light; beautifully white and bright, is it not? It looks so pure and simple, but yet that beam is composed of all sorts of colours mingled together, in such proportions as to form white light. I take a wedge-shaped piece of glass called a prism, and when I apply it in the course of the beam, you see the transformation that has taken place (Fig. 83). Instead of the white light you have now all the colours of the rainbow—red, orange, yellow, green, blue, indigo, violet, marked by their initial letters in the figure. These colours are very beautiful, but they are transient, for the moment we take away

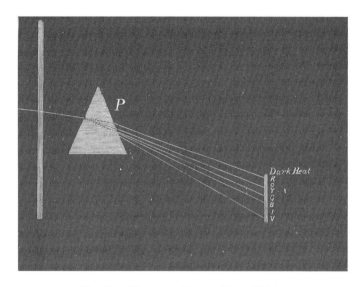

Fig. 83.—How to split up a Ray of Light.

the prism they all unite again to form white light. You see what the prism has done: it has bent all the light in passing through it; but it is more effective in bending the blue light than the red light, and consequently the blue is carried away much further than the red. Such is the way in which we study the composition of a heavenly body. We take a beam of its light, we pass it through a prism, and immediately it is separated into its components; then we compare what we find with the lights given by the different elements, and thus we are enabled to discover the substances which exist in the distant object whose light we have examined. I do not mean to say that the method is a simple one; all I am endeavouring to show is a general outline of the way in which we have discovered the materials present in the stars. The instrument that is employed for this purpose is called the spectroscope. And perhaps you may remember that name by these lines, which I have heard from an astronomical friend:—

"Twinkle, twinkle little star,
Now we find out what you are,

When unto the midnight sky,
We the spectroscope apply."

I am sure it will interest everybody to know that the elements the stars contain are not altogether different from those of which the earth is made. It is true there may be substances in the stars of which we know nothing here; but it is certain that many of the most common elements on the earth are present in the most distant bodies. I shall only mention one, the metal iron. That useful substance has been found in some of the stars which lie at almost incalculable distances from the earth.

THE NEBULÆ.

In drawing towards the close of these lectures I must say a few words about some dim and mysterious objects to which we have not yet alluded. They are what are called nebulae, or little clouds; and they are justly called "little" clouds in one sense, for each of them occupies but a very small spot in the sky as compared with that which would be filled by an ordinary cloud in our air. The nebulae are, however, objects of the most stupendous proportions. Were our earth and thousands of millions of bodies quite as big all put together, they would not be nearly so great as one of these nebulae. Pictures of these objects show them to be like dull patches of light on the background of the sky. Astronomers reckon up the various nebulae by thousands, but I must add that most of them are very small and uninteresting. A nebula is sometimes liable to be mistaken for a comet. The comet is, as I have already explained, at once distinguished by the fact that it is moving and changing its appearance from hour to hour, while scores of years may elapse without changes in the aspect or position of a nebula. The most powerful telescopes are employed in observing these faint objects. I take this opportunity of showing a picture of instruments suitable for such observations. It is the great reflector of the Paris Observatory (Fig. 84).

There are such multitudes of nebulae that I can only show a few of the more remarkable kinds. In Fig. 85 will be seen pictures of a curious object in the constellation of Lyra seen under different telescopic

Fig. 84.—A Great Reflecting Telescope.

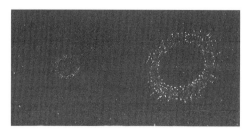

Fig. 85.—The Ring Nebula in Lyra, under different telescopic powers.

powers. This object is a gigantic ring of luminous gas. To judge of the size of this ring let us suppose that a railway were laid across it, and the train you entered at one side was not to stop until it reached the other side, how long do you think this journey would require? I recollect some time ago a picture in *Punch* which showed a train about to start from London to Brighton, and the guard walking up and down announcing to the passengers the alarming fact that "this train stops nowhere." An old gentleman was seen vainly gesticulating out of the window and imploring to be let out ere the frightful journey was commenced. In the nebular railway the passengers would almost require such a warning.

Let the train start at a speed of a mile a minute, you would think, surely, that it must soon cross the ring. But the minutes pass, an hour has elapsed; so the distance must be sixty miles, at all events. The hours creep on into days, the days advance into years, and still the train goes on. The years would lengthen out into centuries, and even when the train had been rushing on for a thousand years with an unabated speed of a mile a minute, the journey would certainly not have been completed. Nor do I venture to say what ages must elapse ere the terminus at the other side of the ring nebula would be reached.

A cluster of stars viewed in a small telescope will often look like a nebula, for the rays of the stars become blended together. A powerful telescope will, however, dispel the illusion and reveal the separate stars. It was, therefore, thought that all the nebulae might be merely clusters so exceedingly remote that our mightiest instruments failed to resolve them into stars. But this is now known not to be the case. Many of these objects are really masses of glowing gas; such are, for instance, the ring nebulae, of which I have just spoken, and the form of which I can simulate by a pretty experiment.

We take a large box with a round hole cut in one face, and a canvas back at the opposite side. I first fill this box with smoke, and there are different ways of doing so. Burning brown paper does not answer well, because the supply of smoke is too irregular and the paper itself is apt to blaze. A little bit of phosphorus set on fire yields copious smoke, but it would be apt to make people cough, and, besides, phosphorus is a dangerous thing to handle incautiously, and I do not want to suggest anything which might be productive of disaster if the experiment were repeated at home. A little wisp of hay, slightly damped and lighted, will safely yield a sufficient supply, and you need not have an elaborate box like this: any kind of old packing-case, or even a band-box, with a duster stretched across its open top and a round hole cut in the bottom, will answer capitally. While I have been speaking my assistant has kindly filled this box with smoke, and in order to have a sufficient supply, and one which shall be as little disagreeable as possible, he has mixed together the fumes of hydrochloric acid and ammonia from two retorts shown in Fig. 86. A still simpler way of doing the same thing is to put a little common salt in a saucer and pour over it a little oil of vitriol; this is put into the box, and over the floor of the box common smelling-salts is to be scattered. You see there are dense volumes of white smoke escaping from every corner of the box. I uncover the opening and give a push to the canvas, and you see a beautiful ring flying across the room, another ring and another follows. If you were near enough to feel the ring you would experience a little puff of wind; I can show this by blowing out a candle which is at the other end of the table. These rings are made by the air which goes into a sort of eddy as it

Fig. 86.—How to make the Smoke-Rings.

passes through the hole. All the smoke does is to render the air visible. The smoke-ring is indeed quite elastic. If we send a second ring hurriedly after the first we can produce a collision, and you see each of the two rings remains unbroken, though both are quivering from the effects of the blow. They are beautifully shown along the beam of the electric lamp, or, better still, along a sunbeam.

We can make many experiments with smoke-rings. Here, for instance, I take an empty box, so far as smoke is concerned, but air-rings can be driven forth from it, though you cannot see them, but you could feel them even at the other side of the room, and they will, as you see, blow out a candle. I can also shoot invisible air-rings at a column of smoke, and when the missile strikes the smoke it produces a little commotion and emerges on the other side, carrying with it enough of the smoke to render itself visible, while the solid black-looking ring of air is seen in the interior. Still more striking is another way of producing these rings, for I charge this box with ammonia, and the rings from it you cannot see. There is a column of the vapour of hydrochloric acid that also you cannot see; but when the invisible ring enters the invisible column, then a sudden union takes place between the vapour of the ammonia and the vapour of the hydrochloric acid: the result is a solid white substance in extremely fine dust which renders the ring instantly visible.

WHAT THE NEBULÆ ARE MADE OF.

There is a fundamental difference between these little rings that I have shown you and the great rings in the heavens. I had to illuminate our smoke with the help of the electric light, for, unless I had done so, you would not have been able to see them. This white substance formed by the union of ammonia and hydrochloric acid has, of course, no more light of its own than a piece of chalk; it requires other light falling upon it to make it visible. Were the ring nebula in Lyra composed of this material we could not see it. The sunlight which illuminates the planets might, of course, illuminate such an object as the ring if it were near us, comparatively speaking; but Lyra is at such a stupendous distance that any light which the sun could send out there would be just as feeble as the light we receive from a fixed star. Should we be able to show our smoke-rings, for instance, if, instead of having the electric light, I merely cut a hole in the ceiling and allowed the feeble twinkle of a star in the Great Bear to shine through? In a similar way the sunbeams would be utterly powerless to effect any illumination of objects in these stellar distances. If the sun were to be extinguished altogether the calamity would no doubt be a very dire one so far as we are concerned, but the effect on the other celestial bodies (moon and planets excepted) would be of the slightest possible description. All the stars of heaven would continue to shine as before. Not a point in one of the constellations would be altered, not a variation in the brightness, not a change in the hue of any star could be noticed. The thousands of nebulae and clusters would be absolutely unaltered; in fact, the total extinction of our sun would be hardly remarked in the newspapers which are, perhaps, published in the Pleiades or in Orion. There might possibly be a little line somewhere in an odd corner to the effect that "Mr. So-and-So, our well-known astronomer, has noticed that a tiny star, inconspicuous to the eye, and absolutely of no importance whatever, has now become invisible."

If, therefore, it be not the sun which lights up this nebula, where else can be the source of its illumination? There can be no other star in the neighbourhood adequate to the purpose, for, of course, such an object would be brilliant to us if it were large enough and bright enough to impart sufficient illumination to the nebula. It would be absurd to say that you could see a man's face illuminated by a candle while the candle itself was too faint or too distant to be visible. The actual facts are, of course, the other way: the candle might be visible, when it was impossible to discern the face which it lighted.

Hence we learn that the ring nebula must shine by some light of its own, and now we have to consider how it can be possible for such material to be self-luminous. The light of a nebula does not seem to be like flame; it can, perhaps, be better represented by the pretty electrical experiment with Geissler's tubes. These are glass vessels of various shapes, and they are all very nearly empty, as you will understand when I tell you the way in which they have been prepared. A little gas was allowed into each tube, and then almost all the gas was taken out again, so that only a mere trace was left. I pass a current of electricity through these tubes, and now you see they are glowing with beautiful colours. The different gases give out lights of different hues, and the optician has exerted his skill so as to make the effect as beautiful as possible. The electricity, in passing through these tubes, heats the gas which they contain, and makes it glow; and just as this gas can, when heated sufficiently, give out light, so does the great nebula, which is a mass of gas poised in space, become visible in virtue of the heat which it contains.

We are not left quite in doubt as to the constitution of these gaseous nebulae, for we can submit their light to the prism in the way I explained when we were speaking of the examination of the stars. Far from us as that ring in Lyra may be, it is interesting to learn that the ingredients from which it is made are not entirely different from substances we know on our earth. The water in this glass, and every drop of water, is formed by the union of two gases, of which one is hydrogen. This is an extremely light material, as you see by a little balloon which ascends so prettily when filled with it. Hydrogen also burns very readily, though the flame is almost invisible. When I blow a jet of oxygen through the hydrogen I produce a little flame capable of giving a very intense heat. For instance, I hold a steel pen in the flame, and it glows and sputters, and falls down in white-hot drops. It is needless to say that as a constituent of water hydrogen is one of the most important elements on this earth. It is, therefore, of interest to learn that hydrogen in some form or other is a constituent of the most distant objects in space that the telescope has revealed to us.

PHOTOGRAPHING THE NEBULÆ.

Of late years we have learned a great deal about nebulae, by the help which photography has given to us. Look at this group of stars which constitutes that beautiful little configuration known as the Pleiades (Fig. 87). It looks like a miniature representation of the Great Bear; in fact, it would be far more appropriate to call the Pleiades the Little Bear than to apply that title to another quite different constel-

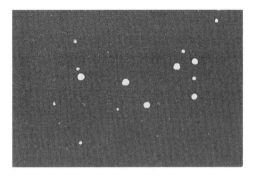

Fig. 87.—The Pleiades.

lation, as has unfortunately been done. The Pleiades form a group containing six or seven stars visible to the ordinary eye, though persons endowed with exceptionally good vision can usually see a few more. In an opera glass the Pleiades becomes a beautiful spectacle; but in a very large telescope the stars appear too far apart to make a really effective cluster. When Mr. Roberts took a photograph of the Pleiades he placed a plate in his telescope, and on that plate the Pleiades engraved their picture with their own light. He left the plate exposed for hours, and on developing it not only were the stars seen, but there were also patches of faint light due to the presence of nebula. It could not be said that the objects on the plate were fallacious, for another photograph was taken, when the same appearances were reproduced.

When we look at that pretty group of stars, which has attracted admiration during all time, we are to think that some of those stars are merely the bright points in a vast nebula, invisible to our eyes or telescopes, though capable of recording its trace on the photographic plate. Does not this give us a greatly increased notion of the extent of the universe, when we reflect that by photography we are enabled to see much which the mightiest of telescopes had previously failed to disclose?

Of all the nebulae, now numbering some thousands, there is but a single one which can be seen without a telescope. It is in the constellation of Andromeda, and on a clear dark night can just be seen with the unaided eye as a faint dull spot. It has happened, before now, that persons noticing this nebula for the first time, have thought they had discovered a comet. I would like you to try and find out this nebula for yourselves.

If you look at it with an opera-glass it appears like a dull spot, rather elongated. You can see more of its structure when you view it in larger instruments, but its nature was never made clear until some beautiful photographs were taken by Mr. Roberts (Fig. 88). Unfortunately, the nebula in Andromedra has not been placed

R. S. Ball

Fig. 88.—Taken from Mr: Roberts's Photograph of the Great Nebula in Andromeda.

conveniently for its portrait. It seems as if it were a rather flat-shaped object, turned nearly edgewise towards us. If you wanted to look at the pattern on a plate, you would naturally hold the plate square in front of you. You would not be able to see the pattern well if the plate were so tilted that the edge was turned towards you. That seems to be nearly the way in which we are forced to view the nebula in Andromeda. We have the same sort of difficulty about Saturn, we are never able to take a square look at his rings. We can trace in the photograph some divisions extending entirely round the nebula, showing that it seems to be formed of a series of rings; and there are some outlying portions which form part of the same system. Truly this is a marvellous object. It is impossible for us to form any conception of the true dimensions of this gigantic nebula, it is so far off that we have never yet been able to determine its distance. Indeed, I may take this opportunity of remarking that no astronomer has yet succeeded in ascertaining with accuracy the distance of any nebula, though everything points to the conclusion that they are at least as far as the stars.

It is almost impossible to apply the methods which we use in finding the distance of a star to the discovery of the distance of the nebulae. These flimsy bodies are usually too ill-defined to admit of being measured with the precision and the delicacy required for the determination of distance. The necessary measurements can only be made from one star-like point to another similar point. If we could choose a star in the nebula and determine its distance, then, of course, we should have the distance of the nebula itself; but the difficulty is that we have, in general, no means of knowing whether the star does actually lie in the nebula. It may, for anything we can tell, lie billions of miles nearer to us than the nebula, or billions of miles further off, and by merely happening to lie in the line of sight, appear to glimmer in the nebula itself.

If we have any assurance that the star is surrounded by nebula, then it may be possible to measure that nebula's distance. It will occasionally happen that grounds can be found for believing that a star which appears to be in the nebula does veritably lie therein, and is not merely seen in the same direction. There are hundreds of stars visible on a good drawing or a good photograph of the Andromeda nebula, and doubtless large numbers of these are merely stars which happen to lie in the same line of sight. The peculiar circumstances attending the history of one star, seem, however, to warrant us in making the assumption that it was certainly in the nebula. The history of this star is a remarkable one. It suddenly kindled from invisibility into brilliancy. How is a change so rapid in the lustre of a star to be accounted for? In a few days its brightness had undergone an extraordinary increase. Of course, this does not tell us for certain that the star lay in the nebula; but the most rational explanation that I have heard offered of this occurrence is that due, I believe, to my friend Mr. Monck. He has suggested that the sudden outbreak in brilliancy might be accounted for on the same principles as those by which we explain the ignition of meteors in our atmosphere. If a dark star, moving along with terrific speed through space, were suddenly to plunge into a dense region of the Great Nebula, heat and light would be evolved sufficiently to transform the star into a brilliant object. If, therefore, we knew the distance of this star at the time it was in Andromeda, we should, of course, learn the distance of the nebula. This has been attempted, and it has thus been proved that the Great Nebula must be very much further from us than is that star of whose distance I attempted some time ago to give you a notion.

We thus realise the enormous size of the Great Nebula. It appears that if, on a map of the nebula, we were to lay down, accurately to scale, a map of the solar system, putting the sun in the centre and all the planets around in their true proportions out to the boundary traced by Neptune, this area, vast as it is, would be a mere speck on the drawing of the object. Our system would have to be enormously bigger than it is if it were to cover anything like the area of the sky included in one of these great objects. Here is a sketch of a nebula (Fig. 89), and on it I have marked a dot which is to indicate our solar system. We may feel confident that the Great Nebula is as big if not bigger than this proportion would indicate.

CONCLUSION.

And now, my young friends, I am drawing near the close of that course of lectures which has occupied us, I hope you will think not unprofitably, for a portion of our Christmas holidays. We have

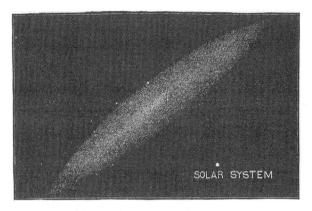

Fig. 89.—To show how small is the Solar System in comparison with a great Nebula.

spoken of the sun and of the moon, of comets and of stars, and I have frequently had occasion to allude to the relative position of our earth in the universe. No doubt it is a noble globe which we inhabit, but I have failed in my purpose if I have not shown you how insignificant is this earth when compared with the vast extent of some of the other bodies that abound in space. We have, however, been endowed with a feeling of curiosity which makes us long to know of things beyond the confines of our own earth. Astronomers can tell us a little, but too often only a little. They will say—That is a star, and That is a planet, and This is so big, and That so far; such is the meagre style of information with which we often have to be content, and, indeed, it is rare for us to receive even so much. The astronomers that live on other worlds, if their faculties be in any degree comparable with ours, must be similarly ignorant with regard to this earth. Inhabitants of our fellow-planets can know hardly anything more than that we live on a globe 8,000 miles across with many clouds around us. Some of the planets would not even pay us the compliment of recognising our existence; while from the otter systems—the countless other systems—of space we are absolutely imperceptible and unknown. Out of all the millions of bodies which we can see, you could very nearly count on your fingers those from which our earth would be visible. This reflection is calculated to show us how vast must be the real extent of that universe around us. Here is our globe, with all its inhabitants, with its great continents, with its oceans, with its empires, its kingdoms, with its arts, its commerce, its literature, its sciences—all of which are naturally of such engrossing importance to us here—and yet it would seem that all these things are absolutely unknown to any inhabitants that may exist elsewhere. I do not think that any reasonable person will doubt that there are inhabitants elsewhere. There are millions of globes, many of them more splendid than ours. Surely it would be presumptuous to say that this is the only one of all the bodies in the universe, on the surface of which life, with all that life involves, is manifested? You will rather think that our globe is but one in the mighty fabric, and that other globes may teem with interest just as ours does. We can, of course, make no conjecture as to what the nature of the life may be elsewhere. Could a traveler visit some other globes and bring back to us specimens of the natural objects that he there found, no collections that the world has ever seen could rival them in interest. When I go into the British Natural History Museum and look around that marvellous collection it awakens in me a feeling of solemnity. I see there the remains of mighty extinct animals which once roamed over this earth; I see there objects which have been dredged from the bottom of the sea at a depth of some miles; there I can examine crystals which have required incalculable ages for their formation; and there I look at meteorites which have travelled from the heavens above down on to the earth beneath. Such sights, and the reflections they awaken, bring before us in an imposing manner the phenomena of our earth, and the extent and interest of its past history. Oliver Wendell Holmes said that the only way to see the British Museum was to take lodgings close by when you were a boy, and to stay in the Museum from nine to five every day until you were an old man: then you would begin to have some notion of what this Institution contains. Think what millions of British Museums would be required were the universe to be adequately illustrated: one museum for the earth, another for Mars, another for Venus— but it would be useless attempting to enumerate them!

Most of us must be content with acquiring the merest shred of information with regard even to our own earth. Perhaps a school-boy will think it fortunate that we are so ignorant with respect to the celestial bodies. What an awful vista of lessons to be learned would open before his view, if only we had a competent knowl-edge of the other globes which surround us in space! I should like to illustrate the extent of the universe by following out this reflec-tion a little further. I shall just ask you to join with me in making a little calculation as to the extent of the lessons you would have to learn if ever astronomers should succeed in discovering some of the things they want to know.

Of course, everybody learns geography and history. We must know the geography of the leading countries of the globe, and we must have some knowledge of their inhabitants and of their gov-ernment, their resources and their civilisation. It would seem shockingly ignorant not to know something about China, or not to have some ideas on the of India or Egypt. The discovery of the New World also involves matters on which every boy and girl has to be instructed. Then, too, languages form, as also we know too well, an immense part of education. Supposing we were so far acquainted with the other globes scattered through space, that we were able to gain some adequate knowledge of their geography and natural his-tory, of the creatures that inhabit them, of their different products and climates, then everybody would be anxious to learn those par-ticulars; and even when the novelty had worn off it would still be right for us to know something about countries perhaps more pop-ulous than China, about nations more opulent than our own, about battles mightier than Waterloo, about animals and plants far stranger than any we have ever dreamt of. An outline of all such matters should, of course, be learned, and as the amount of information would be rather extensive, we will try to condense it as much as possible.

To aid us in realising the full magnificence of that scheme in the heavens of which we form a part, I shall venture to give an illustration. Let us attempt to form some slight conception of the number and of the bulk of the books which would be necessary for conveying an adequate description of that marvellous universe of stars which surround us. These stars being suns, and many of them being brighter and larger than our own sun, it is but reasonable to suppose that they may be attended by planetary systems. I do not say that we have any right to infer that such systems are like ours. It is not improbable that many of the suns around us have a much poorer retinue than that which dignifies our sun. On the other hand, it is just as likely that many of these other suns may be the centres of systems far more brilliant and interesting, with far greater diversity of structure, with far more intensity and variety of life and intelligence than are found in the system of which we form a part. It is only reasonable for us to suppose that, as our earth is an average planet, so our sun is an average star both in size and in the importance of its attendants. We may fairly take the number of stars in the sky at about one hundred millions; and thus we see that the books which are to contain a description of the entire universe — or rather, I should say, of the entire universe that we see—must describe 100,000,000 times as much as is contained in our single system. Of course, we know next to nothing of what the books should contain; but we can form some conjecture of the number of those books, and this is the notion to which I now ask your attention.

So vast is the field of knowledge that has to be traversed, that we should be obliged to compress our descriptions into the narrowest compass. We begin with a description of our earth; nearly all the books in all the libraries that exist at this moment are devoted to matters on or of this earth. They include all branches of history, all languages and literatures and religions, everything relating to life on this globe, to its history in past geological times, to its geography, to its politics, to every variety of manufacture and agriculture, and all the innumerable matters which concern our earth's inhabitants, past and present. But this tremendous body of knowledge must be immensely condensed before it would be small enough to retire to its just position in the great celestial library. I can only allow to the earth one volume of about 500 pages. Everything that has to be said about our earth must be packed within this compass. All terrestrial languages, histories, and sciences that cannot be included between its covers can find no other place on our shelves. I cannot spare any more room. Our celestial library will be big enough, as you shall presently see. I am claiming a good deal for our earth when I regard it as one of the most important bodies in the solar system. Of course it is not the biggest—very far from it; but it seems as if the big planets and the sun were not likely to be inhabited, so that if we allow one other volume to the rest of the solar system, it will perhaps be sufficient, though it must be admitted that Venus, of which we know next to nothing, except that it is as large as the earth, may also be quite as full of life and interest. Mars and Mercury are also among the planets with possible inhabitants. We are, therefore, restricting the importance of the solar system as much as possible, perhaps even too much, by only allowing it two volumes. Within those two volumes every conceivable thing about the entire solar system—sun, planets (great and small), moons, comets, and meteors—must be included, or else it will not be represented at all in the great celestial library. We shall deal on similar principles with the other systems through space. Each of the 100,000,000 stars will have two volumes allotted to it. Within the two volumes devoted to each star we must compress our description of the body itself and of the system which surrounds it;

the planets, their inhabitants, histories, arts, sciences, and all other information. I am not, remember, discussing the contents, but only the number of the books we should have to read ere we could obtain even the merest outline of the true magnificence of the heavens. Let us try to form some estimate as to the kind of library that would be required to accommodate 200,000,000 volumes. I suppose a long straight hall, so lofty that there could be fifty shelves of books on each side. As you enter you look on the right hand and on the left, and you see it packed from floor to ceiling with volumes. We have arranged them according to the constellations. All the shelves in one part contain the volumes relating to the worlds in the Great Bear, while upon the other side may repose ranks upon ranks of volumes relating to the constellation of Orion.

I shall suppose that the volumes are each about an inch and a half thick, and as there are fifty shelves on each side, you will easily see that for each foot of its length the hall will accommodate 800 books. We can make a little calculation as to the length of this library, which, as we walk down through it, stretches out before us in a majestic corridor, with books, books everywhere. Let us continue our stroll, and as we pass by we find the shelves on both sides packed with their thousands of volumes; and we walk on and on, and still see no end to the vista that ever opens before us. In fact, no building that was ever yet constructed would hold this stupendous library. Let the hall begin on the furthest outskirts of the west of London, carry it through the heart of the City, and away to the utmost limits of the east—not a half of the entire books could be accommodated. The mighty corridor would have to be fifty miles long, and to be packed from floor to ceiling with fifty shelves of books on each side, if it is to contain even this very inadequate description of the contents of the visible universe. Imagine the solemn feelings with which we should enter such a library, could it be created by some miracle! As we took down one of the volumes, with what mysterious awe should we open it, and read therein of some vast world which eye had never seen! There we might learn strange problems in philosophy, astonishing developments in natural history; with what breathless interest we should read of inhabitants of an organisation utterly unknown to our merely terrestrial experience! Notwithstanding the vast size of the library, the description of each globe would have to be very scanty. Thus, for instance, in the single book which referred to the earth I suppose just a little chapter might be spared to an island called England, and possibly a page or so to its capital, London. Similarly meagre would have to be the accounts of the other bodies in the universe; and yet, for this most inadequate of abstracts, a library fifty miles long, and lined closely with fifty shelves of books on each side, would be required!

Methuselah lived, we are told, nine hundred and sixty-nine years; but even if he had attained his thousandth birthday he would have had to read about 300 of these books through every day of his life before he accomplished the task of learning even the merest outline about the contents of space.

If, indeed, we were to have a competent knowledge of all these other globes, of all their countries, their geographies, their nations, their climates, their plants, their animals, their sciences, languages, arts, and literatures, it is not a volume, or a score of volumes, that would be required, but thousands of books would have to be devoted to the description of each world alone, just as thousands of volumes have been devoted to the affairs of this earth without exhausting the subjects of interest it presents. Hundreds of thousands of libraries, each as large as the British Museum, would not contain all that should be written, were we to have anything like a detailed description of the universe *which we see*. I specially emphasise the words just written, and I do so because the grandest thought of all, and that thought with which I conclude, brings before us the overwhelming extent of the unseen universe. Our telescopes can, no doubt, carry our vision to an immeasurable distance into the depths of space. But there are, doubtless, stars beyond the reach of our mightiest telescopes. There are stars so remote that they cannot record themselves on the most sensitive of photographic plates.

On the blackboard I draw a little circle, with a piece of chalk. I think of our earth as the centre, and this circle will mark for us the limit to which our greatest telescopes can sound. Every star which we see, or which the photographic plate sees, lies within this circle; but, are there no stars outside? It is true that we can never see them, but it is impossible to believe that space is utterly void and empty where it lies beyond the view of our telescopes. Are we to say that inside this circle stars, worlds, nebulæ, and clusters are crowded, and that outside there is nothing? Everything teaches us that this is not so. We occasionally gain accession to our power by adding perhaps an inch to the diameter of our object glass, or by erecting a telescope in an improved situation on a lofty mountain peak, or by procuring a photographic plate of increased sensibility. It thus happens that we are enabled to extend our vision a little further and to make this circle a little larger, and thus to add a little more to the known inside which has been won from the unknown outside. Whenever this is done we invariably find that the new region thus conquered is also densely filled with stars, with clusters, and with nebulæ; it is thus unreasonable to doubt that the rest of space also contains untold myriads of objects, even though they may never, by any conceivable improvement in our instruments, be brought within the range of our observation. Reflect that this circle is comparatively small with respect to the space outside. It occupied but a small spot on this blackboard, the blackboard itself occupies only a small part of the end of the theatre, while the end of the theatre is an area very small compared with that of London, of England, of the world, of the solar system, of the actual distance of the stars. In a similar way the region of space which is open to our inspection, is an inconceivably small portion of the entire extent of space. The unknown outside is so much larger than the known inside, it is impossible to express the proportion. I write down unity in this corner and a cypher after it to make ten, and six cyphers again to make ten millions, and again, six cyphers more to make ten billions; but I might write six more, aye, I might cover the whole of this blackboard with cyphers, and even then I should not have got a number big enough to express how greatly the extent of the space we cannot see exceeds that of the space we can see. If, therefore, we admit the fact which no reasonable person can doubt, that this outside, this unknown, this unreachable, and, to us, invisible space does really contain worlds and systems as does this small portion of space in which we happen to be placed — then, indeed, we shall begin truly to comprehend the majesty of the universe. What figures are then to express the myriads of stars that should form a suitable population for a space inconceivably greater than that which contains 100,000,000 stars? But our imagination will extend still further. It brings before us these myriads of unseen stars with their associated worlds, it leads us to think that these worlds may be full to the brim with interests as great as those which exist on our world. When we remember that, for an adequate description of the worlds which we can see, one hundred thousand libraries, each greater than any library on earth, would be utterly insufficient, what conception are we to form when we now learn that even this would only amount to a description of an inconceivably small fragment of the entire universe?

Let us conceive that omniscience granted to us an adequate revelation of the ample glories of the heavens, both in that universe which we do see and in that infinitely greater universe which we do not see. Let a full inventory be made of all those innumerable worlds, with descriptions of their features and accounts of their inhabitants and their civilisations, their geology and their natural history, and all the boundless points of interest of every kind which a world in the sense in which we understand it does most naturally possess. Let these things be written every one, then may we say, with truth, that were this whole earth of ours covered with vast buildings, lined from floor to ceiling with book-shelves—were every one of these shelves stored full with volumes, yet, even then this library would be inadequate to receive the books that would be necessary to contain a description of the glories of the sidereal heavens.

Röntgen Light

Silvanus Phillips Thompson*

Röntgen's Discovery—Production of light in vacuum tubes by electric discharges—Exhaustion of air from a tube—Geissler-tube phenomena—The mercurial pump—Crookes's-tube phenomena—Properties of Kathode light—Crookes's shadows—Deflection of Kathode light by a magnet—Luminescent and mechanical effects—Lenard's researches on Kathode rays in air—Röntgen's researches—The discovery of X-rays by the luminescent effect—Shadows on the luminescent screen—Transparency of aluminium—Opacity of heavy metals—Transparency of flesh and leather—Opacity of bones—Absence of reflexion, refraction, and polarisation—Diselectrifying effects of Röntgen rays—Improvements in Röntgen tubes—Speculations on the nature of Röntgen light—Seeing the invisible.

So many erroneous accounts have appeared, chiefly in photographic journals, written by persons unacquainted with physical science, that it seems worth while in beginning a lecture on the subject of Röntgen's rays to state precisely how Röntgen's discovery was made, in the language in which he himself has stated it.

"Will you tell me," asked Mr. H. J. W. Dam in an interview[1] with Prof. Röntgen in his laboratory at Würzburg, "the history of the discovery?"

---

* The original version appeared in: Silvanus Phillips Thompson, Lecture 6, *Light, Visible and Invisible: A Series of Lectures Delivered at the Royal Institution of Great Britain, at Christmas, 1896*, London, 1897, pp. 238–276.
[1] M'Clure's *Magazine*, vol. vi. p. 413.

"There is no history," he said. "I had been for a long time interested in the problem of the kathode rays from a vacuum tube as studied by Hertz and Lenard. I had followed theirs and other researches with great interest, and determined, as soon as I had the time, to make some researches of my own. This time I found at the close of last October [1895]. I had been at work for some days when I discovered something new."

"What was the date?"

"The 8th of November."

"And what was the discovery?"

"I was working with a Crookes's tube covered by a shield of black cardboard. A piece of barium platino-cyanide paper lay on the bench there. I had been passing a current through the tube, and I noticed a peculiar black line across the paper."

"What of that?"

"The effect was one which could only be produced, in ordinary parlance, by the passage of light. No light could come from the tube because the shield which covered it was impervious to any light known, even that of the electric arc."

"And what did you think?"

"I did not think; I investigated. I assumed that the effect must have come from the tube, since its character indicated that it could come from nowhere else. I tested it. In a few minutes there was no doubt about it. Rays were coming from the tube, which had a luminescent effect upon the paper. I tried it successfully at greater and greater distances, even at two metres. It seemed at first a new kind of light. It was clearly something new, something unrecorded."

"Is it light?"

"No." [It can neither be reflected nor refracted.]

"Is it electricity?"

"Not in any known form."

"What is it?"

"I do not know. Having discovered the existence of a new kind of rays, I of course began to investigate what they would do. It soon appeared from tests that the rays had penetrative power to a degree hitherto unknown. They penetrated paper, wood, and cloth with ease, and the thickness of the substance made no perceptible difference, within reasonable limits. The rays passed through all the metals tested, with a facility varying, roughly speaking, [inversely] with the density of the metal. These phenomena I have discussed carefully in my report[2] to the Würzburg Society, and you will find all the technical results therein stated."

Such was Röntgen's own account given by word of mouth. It is entirely borne out by the fuller document, in which in quiet and measured terms Röntgen described to the Würzburg Society his discovery under the title "On a new kind of Rays," and which was the first announcement to the scientific world.

Now you will note that in the whole passage I have read describing the discovery, there is not a word about photography from beginning to end. Photography played no part in the original observation. No photographic plate or sensitised paper was employed. The discovery was made by the use of the luminescent screen, the acquaintance of which you made (if you did not know of it before) at my fourth lecture, when we were dealing with ultra-violet light. On that occasion I showed you a card partly covered with platinocyanide of barium which has been in my possession since 1876. When exposed to invisible ultra-violet light it shone in the dark. No one who has ever used such a luminescent screen can blunder into mistaking it for a photographic plate. Such a screen—a piece of paper covered with the luminescent stuff[3]—was Röntgen using in his investigations. And as luminescent screens are not things to be found lying about by accident, it is evident that its presence on the

---

[2] Ueber eine neue Art von Strahlen (Vorläufige Mittheilung), von Dr. Wilhelm Konrad Röntgen. (Sitzungsberichte der Würzburger physik-medic. Gesellschaft, 1895.)

[3] It is interesting to note that Lenard's investigations of 1894 were conducted by the aid of a luminescent screen composed of paper impregnated with the wax-like chemical called pentadecylparatolylketone.

bench in Röntgen's laboratory on 8th November 1895, when he was deliberately investigating the phenomena observed by Lenard, was in no sense accidental. That you may the better understand the precise nature of Röntgen's discovery, we will repeat the observation with the appliances now at our disposal.

Before you stands a Crookes's tube, which I can at any moment stimulate into activity by passing through it an electric spark from a suitable induction-coil. It shines with visible light, the glass glowing with a beautiful greenish-gold fluorescence. To stop off all visible light, I place over the Crookes's tube this case made of black cardboard, which cuts off not only the visible rays of every sort, but also cuts off the invisible rays of the infra-red and ultra-violet sorts. On the table, just below the tube, lies a sheet of paper covered with platino-cyanide of barium—in fact, a luminescent screen. And, on passing the electric discharge through the shielded Crookes's tube you will all see that this luminescent sheet at once shines in the dark; while across it—as those who are near may observe—there falls obliquely a dark line which is simply a shadow of a small support that stands between the tube and the screen. Something evidently is causing that sheet of luminescent paper to light up. Can the effect come from anywhere else than from the tube? Try by interposing things, and see whether they cast shadows on the paper. The nearest thing at hand is a wooden bobbin, on which wire is wound. If I interpose it, it casts a shadow on the paper. But looking at the shadow one notices, curiously enough, that while the wire casts a decided shadow, the wood casts scarcely any. I hold up the screen that you may see the shadow more plainly. Yes! there is something coming from that tube which causes the screen to light up, and which casts on the screen shadows of things held between tube and screen. This light—if light it be—comes from the tube. But is it light? Light, as we know it, cannot pass through black cardboard. If it be light it is light of some wholly new and more penetrative kind. I move away, still holding the screen in my hand, to greater distances. Here, two metres away, the screen still shines, though less brilliantly. And, note, it shines whether its face or its back be presented toward the tube. The rays, having penetrated the shield of black cardboard that encloses the tube, can also penetrate the paper screen from the back, and make the chemically-prepared face shine. Let us follow Röntgen farther as he investigated the penetrative power of the rays. I interpose a block of wood against which a pair of scissors has been fixed by nails. You can see on the screen the shadow of the scissors; the light passes through the wood, though not so brightly, for the wood intercepts some of the rays. Paper, cardboard, and cloth are easily penetrated by them. The metals generally are more opaque than any organic substance, and they differ widely amongst one another in their transparency. Thin metal foil of all kinds is more or less transparent; but when one tries thicker pieces they are of different degrees of opacity. Ordinary coins are opaque. A golden sovereign, a silver shilling, and a copper farthing are all opaque, but the lighter metals such as tin, magnesium, and aluminium, notably the latter, are fairly transparent. Here is my purse of leather with a metal frame. I have but to hold it between the tube and the screen to see its contents—two coins and a ring—for leather is transparent to these rays. A sheet of aluminium about the twentieth of an inch thick, though opaque to every other previously-known kind of light is for this kind of light practically transparent. On the other hand lead is very opaque. Röntgen found opacity to go approximately in proportion to density. It is now found that those metals which are of the greatest atomic weight are the most opaque to Röntgen light. Uranium, the atomic weight of which is 240, is the most opaque; whilst lithium whose atomic weight is only 7, and which will readily float on water, is exceedingly transparent. In fact I have never yet got a good shadow from lithium. This relation extends not only to the metals themselves but to their compounds. Thus the chloride of lithium is more transparent than the chloride of zinc or than the chloride of silver. Finding that the denser constituents were the more opaque, and that while glass and stone are tolerably opaque such substances as gelatine and leather were comparatively transparent, it occurred to Röntgen that bone would be more opaque than flesh—and so it proved: for interposing the hand between the tube and the screen we find that while the flesh casts a faint shadow the bones cast a much darker one, and so we are able to see upon the luminescent screen, in the darkness, the shadow of the bones of the hand, and of the arm. This is truly seeing the invisible.

But now the investigation took another turn. So far there has been no mention of photography. But the peculiar penetrative light having been discovered, and the shadows having been seen on the luminescent screen, it was a pretty obvious step to register these shadows photographically. For, as was already well known in the case of ultra-violet light, the rays that stimulate fluorescence and phosphorescence are just those rays which are most active chemically and photographically. Hence it was to be expected that these new rays would affect a photographic plate. This Röntgen proceeded to verify. He obtained a photograph of a set of metal weights that were shut up in a wooden box. Also of a compass, showing the needle and dial through the thin brass cover. He then put his tube under a wooden-topped table; and laying his hand on the table above it, and poising over it a photographic dry-plate, face downwards, he threw upon the plate, by light which passed upwards through the table top, a shadow of his hand. So for the first time he succeeded in photographing the bones of a living hand. It was the photography of the invisible. But, note, even here there is no "new photography." The only photography in the matter is the well-known old photography of the dry-plate, which must first be exposed and afterwards developed in the dark-room.

And now, though it anticipates somewhat the course of this lecture, since the process of photographic development in the dark-room requires a little time, I will proceed to take a few photographs which will then be taken to the dark-room to be developed, and will afterwards be brought back and shown you upon the screen by means of the lantern.

[In the experiments which followed photographs were taken of the hands of a boy and of a girl, also shadows cast by sundry gems, including a fine Burmese ruby, a sham ruby, a Cape diamond, and an Indian diamond.]

Retracing our steps in the order of discovery I must at once take you back, nearly two hundred years, to the time of Francis Hauksbee, when, with the newly invented electric machine, and the newly perfected air-pump, the first experiments were made on the peculiar light produced by passing an electric spark into a partial vacuum. About that time Europe was nearly as much excited—considering the state of knowledge and the slow means of communication—over the "mercurial phosphorus," as it was last year over the Röntgen rays. This "mercurial phosphorus" was simply a little glass tube, such as that (Fig. 130) which I hold in my hand. It contains a few drops of quicksilver; and the air that otherwise would fill the tube has been mostly pumped out by an air-pump, leaving a partial vacuum. I have but to shake the tube and it flashes brightly with a greenish light. The friction of the mercury against the glass walls sets up electric discharges, which flash through the residual air, illuminating it at every motion.

While I have been talking to you an air-pump in the basement, driven by a gas-engine, has been at work exhausting a large oval-shaped glass tube. Only perhaps $\frac{1}{300}$ part of the air originally in it

FIG. 130.

remains. On sending through it from top to bottom the sparks from an induction coil, it is filled with a lovely pale crimson glow, which changes at the lower end to a violet-coloured tint. On reversing the connections so as to send the discharge upwards the violet-coloured part is seen at the top. It shows you, in fact, the end at which the electric discharge is leaving the tube. The pale glow of this primitive vacuum tube is rich in light of the ultra-violet kind, which, as you know, readily excites fluorescence. I have but to hold near it my platino-cyanide screen for you to observe the rich green fluorescence. My hand will cast a shadow on the screen if I interpose it, but there are no bones to be seen in the shadow. For here there is none of the penetrative Röntgen light: the fluorescence is due to ordinary ultra-violet waves, to which flesh and cardboard are quite opaque. If the tap is turned on to readmit the air you see how the rosy glow contracts first into a narrowing band, then into a mere line, which finally changes into a flickering forked spark of miniature lightning; and all is over until and unless we pump out the air again. Another beautiful effect is shown by use of an exhausted glass jar, within which is placed a cup of uranium glass, as described fifty years ago by Gassiot. The discharge overflows the cup in lovely streams of violet colour, while the cup itself glows with vivid green fluorescence. Some thirty years ago vacuum tubes became an article of commerce, and were made in many complex and beautiful shapes by the skill of Dr. Geissler of Bonn, who devised a form of mercurial air-pump[4] for the purpose of extracting the air more perfectly; though the degree of vacuum, which sufficed to display the most brilliant colours when stimulated by an electric discharge, is far short of that which is requisite in the modern vacuum tubes of which I must presently speak. Here is a Geissler's tube showing wondrous effects when the spark discharge

---

[4] See the author's monograph on *The Development of the Mercurial Air-Pump*, published in 1888, by Messrs. E. and F. N. Spon.

FIG. 131.

is passed into it. Strange flickering striations palpitate along the windings of the glass tubes which themselves glow with characteristic fluorescence. Soda-glass fluoresces with the golden-green tint, lead glass with a fine blue, and uranium glass with a brilliant green. The violet glow which appears in the bulb at one end of the tube surrounds the metal terminal by which the current leaves the tube, and is itself due to nitrogen in the residual air. Each kind of gas gives its own characteristic tint. And with any kind of gas within the tube the luminous phenomena are different at different degrees of exhaustion.

I have here a set of eight tubes, all of the same simple shape (Fig. 131), but they differ in respect of the degree of vacuum within them. Platinum wires have been sealed through the ends of each, one wire a to serve as the *anode* or place where the electric current enters, another wire k to serve as *kathode* or place where the current makes its exit from the tube. Both anode and kathode are tipped with aluminium, as this metal does not volatilise so readily under the electric discharge. The small side-tube s by which the tube was attached to the pump during exhaustion is hermetically sealed to prevent air from re-entering. The first tube of the set is full of air at ordinary pressure, and does not light up at all. The length between anode and kathode (about 12 inches) is so great that no spark will jump between them. In the second tube the air has been so far pumped away that only about $\frac{1}{5}$ of the original air remains. Across this imperfect vacuum forked brush-like bluish sparks dart. The third tube has been exhausted to about $\frac{1}{20}$ part; that is to say, $\frac{19}{20}$ of the air have been removed. It shows, instead of the darting sparks, a single thin red line, which is flexible like a luminous thread. In the fourth tube the residual air is reduced to $\frac{1}{40}$ or $\frac{1}{50}$ part; and you note that the red line has widened out into a luminous band from pole to pole, while a violet mantle makes its appearance at each end, though brighter at the kathode. In the fifth tube, where the exhaustion has been carried to about $\frac{1}{500}$, the luminous column, which fills the tube from side to side, has broken up into a number of transverse striations which flicker and dance; the violet mantle around the kathode has grown larger and more distinct. It has separated itself by a dark space from the flickering red column, and is itself separated from the metal kathode by a narrow dark space. The degree of exhaustion has been carried in the sixth tube to about $\frac{1}{10000}$: and now the flickering striations have changed both shape and colour. They are fewer, and whiter. The light at the anode has dwindled to a mere star; whilst the violet glow around the kathode has expanded, and now fills the whole of that end of the tube. The dark space between it and the metal kathode has grown wider, and now the kathode itself exhibits an inner mantle of a foxy colour, making it seem to be dull and hot. The glass, also, of the tube shows a tendency to emit a green fluorescent light at the kathode end. In the seventh tube the exhaustion has been pushed still farther, only about $\frac{1}{50000}$ part of the original air being left. The luminous column has subsided into a few greyish-white nebulous patches. The dark space around the kathode has much expanded, and the glass of the tube exhibits a yellow-green fluorescence. In the eighth tube only one or two millionths of the original air are present; and it is now found much more difficult to pass a spark through the tube. All the internal flickering clouds and striations in the residual gas have disappeared. The tube looks as if it were quite empty: but the glass walls shine brightly with the fine golden-green fluorescence, particularly all around the kathode. If we had pushed the exhaustion still farther, the internal resistance would have increased so much that the spark from the induction coil would have been unable to penetrate across the space from anode to kathode.

To attain such high degrees of exhaustion as those of the latter few tubes recourse must be had to mercurial air-pumps; no mechanical pump being adequate to produce sufficiently perfect vacua. The Sprengel pump, invented in 1865 by Dr. Hermann Sprengel, is an admirable instrument for the purpose. But it was modified and greatly improved[5] about 1874 by Mr. Crookes, whose form of pump is shown in Fig. 132. Mercury is placed in a supply-vessel, which can be raised to drive the mercury through the pump, and lowered, when empty, to be refilled. This vessel is connected by a flexible indiarubber tube to the pump, which consists of glass-tubes fused together. From the pump-head the mercury falls in drops down a narrower tube, called the fall-tube, and each drop as it falls acts as a little piston to push the air in front of it, and so gradually to empty the space in the farther part of the tube. A drying-tube, filled with phosphoric acid to absorb moisture, is interposed between the pump and the vacuum-tube that is to be exhausted. It is usual to add a barometric gauge to show the degree of vacuum that has been reached.

Before you, fixed against the wall, is a mercury-pump substantially like Fig. 132, but having three fall-tubes instead of one, so as to work more rapidly. Through these fall-tubes mercury is dropping freely; the pump being at the present moment employed in

---

[5] These improvements comprised the following:—A method of lowering the supply-vessel to refill it with the mercury that had run through the pump; the use of taps made wholly of platinum to ensure tightness; the use of a spark-gauge to test the perfection of the vacuum by observing the nature of an electric spark in it; the use of an air-trap in the tube leading up to the pump-head; the method of connecting the pump with the object to be exhausted, by means of a thin, flexible, spiral glass tube; the method of cleansing the fall-tube by letting in a little strong sulphuric acid through a stoppered valve in the head of the pump. In carrying out these developments Mr. Crookes was assisted by the late Mr. C. Gimingham, whose later contributions to the subject are described in the author's monograph on the Mercurial Air-pump.

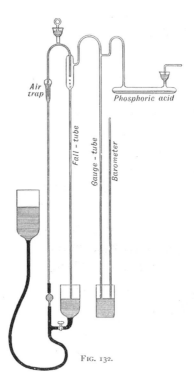

Fig. 132.

the exhaustion of a Crookes's tube, which has been sealed to it by a narrow glass tube. When the exhaustion has been carried far enough, this narrow pipe will be melted with a blow-pipe, so as to seal up the tube and enable it to be removed from the pump.

It was with such a pump as this that Crookes was working from 1874 to 1875 in the memorable researches on the repulsion caused by radiation, which culminated in the invention of that exceedingly beautiful apparatus the *radiometer*, or light-mill, which we were using in my last lecture. From that series of researches Mr. Crookes was led on to another upon the phenomena of electric discharge in high vacua. Professor Hittorf of Münster had already done some excellent work in this direction. He had noted the golden-green fluorescence around the kathode when the exhaustion was pushed to

Fig. 133.—Professor William Crookes, F.R.S.

a high degree; and he had found that this golden glow, unlike the luminous column which at a lower exhaustion fills the vacuous tube, refuses to go round a corner. He had even found that it could cast shadows, owing to its propagation in straight lines.

Starting at this point on his famous research, Crookes investigated the properties of this kathode light, and found it to differ entirely from any known kind of radiation. It appeared to start off from the surface of the kathode and to move in straight lines, penetrating to a definite distance, the limit of which was marked by the termination of the "dark space," according to the degree of exhaustion, and causing the bright fluorescence when the exhaustion was carried so far that the dark space expanded to touch the walls. Acting on this hint he proceeded to construct tubes in which the

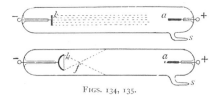

FIGS. 134, 135.

kathode, instead of being as previously a simple wire, was shaped as a flat disk, or as a cup (Figs. 134, 135). From the flat disk the kathode rays streamed backwards in a parallel beam. Crookes regarded these kathode streams as flights of negatively-electrified molecules shot backwards from the metal surface. Doubtless such flying molecules of residual gas there are; and they take part in the phenomenon of discharge, bombarding against the opposite wall of the tube. There are, however, strong reasons for thinking that the kathode rays are not merely flights of "radiant matter," but that the flying molecules are accompanied by ether-waves or ether-motions which cause the fluorescence on the walls of the tube. Be that as it may, Crookes found the kathode rays to be possessed of several remarkable properties. Not only could they excite fluorescence and phosphorescence to a degree previously unknown, but they exercised a mechanical force against the surfaces on which they impinged. They cast shadows of objects interposed in their path; and were capable of being drawn aside by the influence of a magnet, just as if they were electric currents.

Here are some Crookes's tubes which display the luminescent effects. At the top of the first is a small flat disk of aluminium to serve as kathode. From it shoots downward a kathode-beam upon a few Burmese rubies fixed below. They glow with a crimson tint more intense than if they had been red-hot. In a similar tube is a beautiful phenakite,[6] looking like a large diamond. When exposed to the

---

[6] A species of white emerald found in the Siberian emerald mines, and often sold in Russia as a Siberian diamond. It is not so brilliant as a diamond, though much more rarely met with.

kathode rays it luminesces with a lovely pale blue tint. In the third is placed a common whelk shell, which has been lightly calcined. As the kathode rays stream down upon it it lights up brilliantly. And, after the electric discharge has been switched off, the shell continues for some minutes to phosphoresce with a persistent glow.

In the next tube, which contains a sheet of mica painted with a coat of sulphate of lime so that you may better see the bright trace of its luminescence, a narrow kathode ray is admitted through a slit at the bottom, and extends in a fine bright line upwards. Holding a magnet near it, I draw the kathode ray on one side, illustrating its deflectibility.

To illustrate the mechanical effect of the kathode rays I take a Crookes's tube, having at its ends flat disks of metal as electrodes. Between them is a nicely-balanced paddle-wheel, the axle of which runs upon a sort of little railway. On sending the spark from the induction-coil through the tube the little wheel is driven round and runs along the rails. Its paddles are driven as if a blast issued from the disk which serves as kathode. On reversing the current its motion is reversed.

Here (Fig. 136) is a Crookes's tube of a pear shape, having a piece of sheet-metal in the form of a Maltese cross set in the path of the kathode rays. See what a fine shadow the cross casts against the broad end of the tube; for the whole end of the tube glows with the characteristic golden-green luminescence, except where it is shielded from the rays by the metal cross.

FIG. 136.

With this tube I am able to show you a most interesting and novel experiment discovered only a few days ago by Professor Fleming. If you surround the tube with a magnetising coil through which an electric current is passed, the magnetic field produces a remarkable effect on the shadow. Instead of pulling it on one side (as a horse-shoe magnet would do), the magnetising coil causes the cross to rotate on itself, and at the same time to grow smaller. To show the effect more conveniently I have put the magnetising coil not around the tube itself, but around an iron core beyond the end of the tube. So I am able to diminish or augment the effect by simply moving the tube away from the iron core, or by bringing it nearer. As I move it up, the shadow of the cross contracts, and grows smaller but brighter. It also twists round and turns completely over top for bottom as it vanishes into a mere point. But just as it vanishes you see its place taken by a second large shadow, which, as I push the tube still closer to the magnetised core, grows brighter and also turns round and contracts as its predecessor did. Its arms are more curved than those of the first cross. At the same moment when the second shadow-cross appears a third shadow makes its appearance as a distorted annular form against the walls of the tube between the metal cross and the kathode. Its position is such that the shadow seems to have been cast as by rays diverging from the other end of the tube. As yet we know not the explanation of these remarkable facts.

The last tube of this set that illustrates Crookes's researches has as kathode a large hollow cup of aluminium at the bottom (Fig. 137). This concave kathode focuses the kathode rays by converging them to a point in space a little above the centre of the tube. Crookes found that if the kathode rays were in this way focused upon anything, they produced great heat. Glass was melted, diamonds charred, platinum foil heated red-hot and even fused by the impact of the concentrated kathode stream. In the focusing-tube now before you—an old one, made more than ten years ago—there is a piece of thin platinum foil hung in the tube to be heated by the rays. But it has become displaced and no longer hangs in the focus. Yet

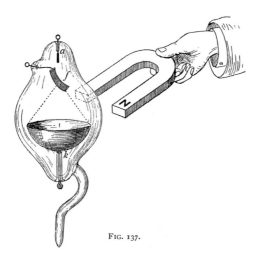

Fig. 137.

by holding a small horse-shoe magnet outside the tube to deflect the rays a little, I can displace the focus until it falls upon the surface of the platinum foil, which you now see is raised to a bright red heat.

Since the date, now nearly twenty years ago, when these most beautiful and astonishing observations were made by Crookes, there has been much speculation as to the nature of these interior kathode rays; their properties were so extraordinarily different from anything in the nature of ordinary light that even the name "ray" as applied to them seemed out of place. Crookes's own term, "radiant matter," was objected to as necessarily implying their material nature; and yet no other explanation of them seemed reasonable than Crookes's own suggestion that they consisted of flights of electrified particles. It was supposed that they could only exist in a vacuum tube under an exceedingly high condition of exhaustion.

However in 1894 Dr. Philipp Lenard, acting on a hint afforded by an observation of Professor Hertz[7] succeeded in bringing out

---

[7] Hertz noticed that when a very thin metal film was interposed inside the Crookes tube, the glass still fluoresced under the kathode discharge. He found this still to be the case when the film was replaced by a piece of thin aluminium foil which was quite opaque to light.

S. P. Thompson

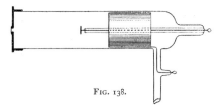

FIG. 138.

the kathode rays into the air at ordinary pressure. For this purpose he fitted up a tube with a small window of thin aluminium
foil opposite the kathode, as shown in Fig. 138. The general form
of tube was the same as that previously used by Hertz, namely,
cylindrical, with a small kathode disk on the end of a central wire,
protected by an inner glass tube. The anode was a cylindrical
metal tube surrounding the kathode. Upon the further end of the
tube was cemented a brass cap, having at its middle a small hole
covered with aluminium foil $\frac{1}{10000}$ inch thick. Through this "window," when the tube was highly exhausted, there came out into
the open air rays which, if not actual prolongations of the kathode rays, are closely identified with them. They can be deflected
by a magnet—though in varying degrees depending on the internal vacuum. They can excite luminescence. Lenard explored them
by using a small luminescent screen of paper covered with a
chemical called pentadecylparatolylketone. He found them to
be capable of affecting a photographic dry-plate; and studied
both by the luminescent screen and by the photographic plate
their power of penetrating materials. He found that air at ordinary pressure was not very transparent, acting toward them as a
turbid medium. He found them to pass through thin sheets of
aluminium and even of copper. He also caused them to affect a
photographic plate that was completely enclosed in an aluminium
case, and to discharge an electroscope enclosed in a metal box. All
this work was done in 1894 and 1895 and duly published. Though

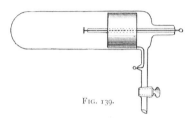

Fig. 139.

it excited no public notice, it was regarded by physicists as of very great importance.

As you were told at the beginning in Röntgen's own account of the matter, his research began with the deliberate aim of reinvestigating the problem of the emission of kathode rays from the vacuum tube as studied by Hertz and Lenard. So as Lenard had done, he employed a luminescent screen to explore the rays, and used a Crookes tube (Fig. 139) of a form closely resembling Lenard's, and indeed identical with that previously employed by Hertz. The end opposite the kathode was simply of glass, without any brass cap or aluminium window. Thus prepared he found what I have already described, those mysterious rays which with characteristic modesty he described as "X-rays," but which will always be best known as Röntgen's rays. They are not kathode rays, though caused by them. Kathode rays will not pass through glass, and are deflected by a magnet. Röntgen rays will pass through glass and are not deflected by a magnet. They seem indeed to be formed by the destruction of the kathode rays, having for their origin the spot where the kathode rays strike against any solid object, best of all against some heavy metal such as platinum or uranium. Neither are they ordinary light of either the infra-red or of the ultra-violet kind, though they resemble the latter in their chemical activity and in so freely exciting luminescence. But ultra-violet light can, as we have seen in previous lectures, be reflected, refracted, and polarised, while Röntgen light cannot.[8] Nor are Röntgen's rays the same thing as Lenard's rays; for the latter are in various degrees deflectible by the magnet; and air is toward them relatively much more opaque than it is for Röntgen's rays. Röntgen seems to have been fortunate in having the means of producing the most perfect exhaustion by his vacuum pump: for on the perfection of the vacuum more than on any other detail does the successful production of the Röntgen rays depend. The vacuum, which is abundantly good enough to evoke luminescence, or to show the shadow of the cross, or to produce the heating at the focus, or to drive the "molecule mill," does not suffice to generate the Röntgen rays. For this last purpose the exhaustion must be carried to a higher point—to a point so high indeed that the tube is on the verge of becoming non-conductive.

Here let me say a word about the man himself and his material surroundings. Still in the prime of life, at the age of fifty-one, Professor Wilhelm Konrad Röntgen had already made himself a

---

[8] Reflexion there is, but not of a regular kind; the supposed cases of true reflexion announced by Lord Blythswood and others belong to the category of myths. There is diffuse reflexion of Röntgen rays from polished metals, particularly from zinc, just as there is diffuse reflexion of ordinary light from white paper. As to refraction, Perrin in Paris, and Winkelmann in Jena, have independently found what they think evidence of feeble refraction through aluminium prisms. But the deviation (which is towards the refracting edge) is so excessively small as to be scarcely distinguishable from mere instrumental errors. Polarisation has been looked for by many skilled observers, using many materials including tourmalines. Only one success has been alleged, by MM. Galitzine and Karnojitzky, using tourmaline; but their result has not been confirmed and is probably erroneous. Neither has interference of Röntgen light yet been shown to be possible. Several observers have professed that they have obtained diffraction fringes from which the wave-length of the Röntgen rays could be measured. But some of these measurements show a wave-length greater than that of red light, and others less than that of ordinary ultra-violet: they are probably all due to some unnoticed source of error. None of them can be accepted without subsequent confirmation by other experimenters, and this is not yet forthcoming.

name among physicists by his work in optics and electricity before the date of the brilliant discovery that gave him wider fame. He occupies the chair of Physics in the University of Würzburg in Bavaria, and lives and works in the physical laboratory of the University. The little town of Würzburg, of 61,000 inhabitants, boasts a university frequented by 1490 students, and supported with an income of £41,000 a year, of which more than half is contributed by the State. There are 53 professors and 40 assistants. Its buildings comprise a group of laboratories and institutes devoted to chemistry, physiology, pathology, mineralogy, and the like. Its physical laboratory, a neat detached block of buildings, wherein also the professor has his residence, is of modern design. Its equipment for the purpose of research is infinitely better than that of the University of London;[9] and it is expected of the professor that he shall contribute to the

---

[9]From a Report recently presented to the Convocation of the University of London, it appears that the physical and chemical laboratories of the University are practically non-existent. "There are three rooms at Burlington House which are occasionally used as laboratories during examinations, and for examinational purposes only. The largest of these is a large hall lit from the top. When used as a chemical laboratory, it is fitted up with working benches down the middle and along the two sides, the benches being divided into separate stalls to isolate candidates in their work. It was stated that the middle stalls and benches are taken down when the hall is used for written examinations, and are re-erected when a chemical examination is to be held. In a second hall, also lighted from above, where frequent written examinations are held, temporary arrangements are made whenever an examination in practical physics is to be held. A curtain of black cloth slung across one end of the room gave partial obscurity over the tables where photometric and spectroscopic apparatus was placed. The third room, sometimes called the galvanometer room, is a smaller room in the basement, artificially lighted, and used chiefly for printing, except at the times of examinations in practical physics." Such is the melancholy state of things in a University where everything is sacrificed on the altar of competitive examinations.

Bavaria has a population of 6,000,000. It supports the three Universities of Munich, Erlangen, and Würzburg, with a total of over 6000 students, at

FIG. 140.—PROFESSOR W. K. RÖNTGEN.

advancement of science by original investigations. With such material and intellectual encouragements to research as surround the university professor in even the smallest of universities in Germany, what wonder that advances are made in science? Would that a like stimulus were existent in England. The Professor of Physics in the

---

a cost of £150,000 a year, of which £93,000 is provided by the State. London, with a population of 5,000,000, has the University of London, a mere Examining Board, to which come up for intermediate and degree examinations about 2000 students yearly, of whom a large proportion are from the provinces. It has no professors. Its laboratories are in the deplorable position above mentioned. So far from being endowed by the State, it pays in to the State about £16,500 a year, and nominally receives back about £16,280 as a parliamentary grant. It receives no subvention from the municipality. Its library is closed for a large portion of the year, the room being used for examination purposes almost every day.

University of London has made no discovery like that of Professor Röntgen, for the very good reason that the University of London has neither appointed any Professor of Physics, nor built any physical laboratory where he might work. Neither the State nor the municipality has provided it with the necessary funds. Its charter precludes it from doing anything for science except hold examinations! Perhaps some day London may have a university worthy of being mentioned beside that of Würzburg, which is eleventh only in size amongst the universities of Germany.

Röntgen had so thoroughly explored the properties of the new rays by the time when his discovery was announced, that there remained little for others to do beyond elaborating his work. One point deserves notice; namely, the improvement of the tubes. Röntgen held the view that his rays originated at the fluorescent spot where the kathode rays struck the glass. This led some persons to the idea that fluorescence was advantageous. Several workers, however, discovered about the same time that if the kathode rays were focused upon a piece of metal the emission of Röntgen light became more copious. When studying early last year the conditions under which the rays were produced, I found that the best radiators are substances which do not fluoresce—namely, metals. I found zinc, magnesium, aluminium, copper, and iron to answer; but platinum was better than these, and uranium best of all. Directing the kathode discharge against a target or "antikathode" of platinum fixed in the middle of the tube, I carefully watched, by aid of a luminescent screen, the emissive activity of the surface during the process of exhaustion. After the stage of exhaustion has been reached at which Crookes's shadows are produced, one must go on further exhausting before any trace of Röntgen rays appear. The first luminosity seems to come (as in Fig. 141) from both front and back of the target at once; an oblique line, corresponding to the plane of the "antikathode" or target of metal, being seen on the screen between two partially luminescent regions. On continuing

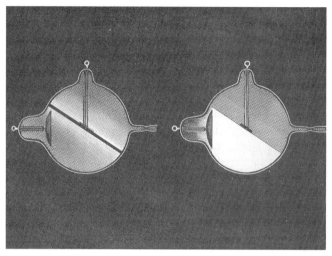

FIG. 141.   FIG. 142.

the exhaustion, the light behind dies out while that in front increases, as in Fig. 142, the rays being emitted copiously right up to the plane of the antikathode. This lateral emission is quite unlike anything in the emission or reflexion of ordinary light, and has to be accounted for in any theory of the Röntgen rays. I have myself observed[10] that within the tube there are some other rays given off in a similar way, along with the Röntgen rays, but which are not Röntgen rays, for they can be deflected by a magnet, and more nearly resemble the kathode rays. It is these that produce on the glass wall of the tube a well-marked fluorescence delimited (as in Fig. 142) by an oblique plane corresponding to the delimitation of Röntgen rays seen in the fluorescent screen. The tube which I used at the beginning of this lecture, and which we will use again at the close of the lecture to show you your own bones, is of the focus type (Fig. 143). It is of the pattern devised by Mr. Herbert Jackson, of King's College. The concentration of the kathode rays

---

[10] See *Electrician*.

FIG. 143.

upon the little target of platinum (which often becomes red-hot) has the advantage not only of allowing a more copious emission of Röntgen rays than would be possible if the antikathodal surface were the glass wall, but also of causing the Röntgen rays to issue from a small and definite source so that the shadows cast by objects are more sharply defined. Here are two still more recent tubes (Figs. 144, 145) constructed for me by Mr. Böhm, in which the focus principle is preserved; but in which there is the improvement that the antikathode T is not used also as an anode. It is an insulated target of platinum, while the anodes are aluminium rings through which the cone of kathode rays passes. These tubes are not liable to blacken, as is the case with tubes in which the antikathode is also used as anode. The tube (Fig. 145) has two concave electrodes, either or both of which may be used as kathode; it is a convenient form for those cases in which an alternating current is employed.

In another direction many efforts have been made at improvement of the luminescent screen. At first good barium platino-cyanide was not to be procured, and hydrated potassium platino-cyanide was

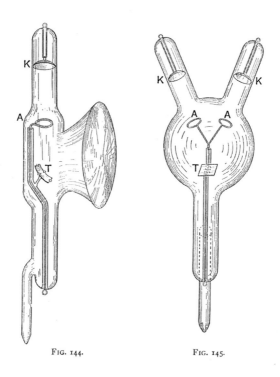

Fig. 144.   Fig. 145.

found far superior. But the good barium salt now procurable is quite as luminescent, and is less troublesome to manage. One result of the ignorance which at first prevailed as to the real origin of Röntgen's discovery was that various experimenters up and down the world supposed themselves to have invented something when they took to using fluorescent screens. One man puts a fluorescent screen at the bottom of a pasteboard tube, with a peep-hole lens at the top, and calls it a "cryptoscope." Another, in another part of the globe, puts a fluorescent screen at the bottom of a nice cardboard box furnished with a handle and a flexible aperture to fit to the eyes, and styles it a "fluorescope." Both are useful; but the only invention in the whole thing is Röntgen's.

Within a few days of the publication of Röntgen's discovery another effect, however, which had escaped Röntgen's scrutiny, was observed by several independent observers. It had been known for several years that when ultra-violet light falls upon an electrically-charged surface it will cause a diselectrification, but only if the surface is negatively charged. Ultra-violet light will not diselectrify a positive charge.[11] But Röntgen rays are found to produce a diselectrification of a metal surface (in air) whether the charge be positive or negative. Here is a convenient arrangement for exhibiting the experiment. An electroscope made on Exner's plan with three leaves—the central one a stiff plate of metal—is charged, and then exposed to Röntgen light. The three leaves are made of aluminium, aluminium foil being better than leaf-gold for electroscopes. They are supported within a thin flask of Bohemian glass entirely enclosed, except at the top, in a mantle of transparent metallic gauze. After the leaves have been charged—either positively by a rod of rubbed glass, or negatively by a rod of rubbed celluloid—a metal cap is placed over the top (Fig. 146).

The leaves, being thus completely surrounded by metal, are effectually screened from all external electrical influences. My

FIG. 146.

---

[11] Ultra-violet light will not diselectrify a *metal* surface in air unless that surface is negatively charged. I have observed a case, however, in which a positively-electrified body was discharged by ultra-violet light, but it was not a metal surface, nor in air.

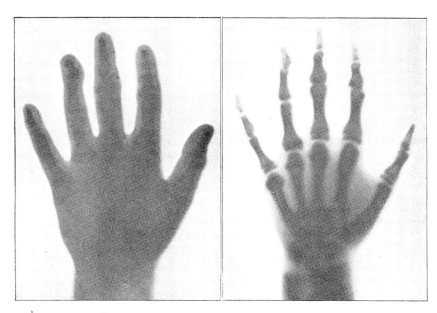

FIG. 147.   FIG. 148.

electroscope is now charged. To enable you to see the effect better, a beam of light is directed upon it, throwing a magnified shadow of the leaves upon the white screen. Then, exposing the electroscope to Röntgen light from a focus tube situated some 18 inches away, you see the leaves at once closing together, proving the diselectrification. It succeeds whether the charge be positive or negative in sign.

It now only remains for me to exhibit to you the photographs which were taken at the beginning of this lecture, and a number of others prepared as lantern-slides. In Figs. 147, 148 we have the hand of a poor child aged thirteen, a patient in St. Bartholomew's Hospital. She was brought to my laboratory that the deformities of her hands might be examined. The first of the two plates was insufficiently exposed, with the result that the bones scarcely show through the flesh at all. The second plate was over-exposed, and the rays have penetrated the flesh so thoroughly that only the bones appear.

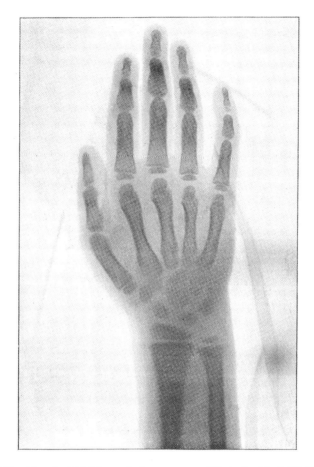

FIG. 149.—Hand of Child, aged eleven years. (*Photo. by Mr. J. W. Gifford.*)

Fig. 149 is the hand of a child of eleven years old. In a child's hand the bones are not yet completely ossified, their ends being still gelatinous and transparent, so that there seem to be gaps between them. Compare this with the hand of a full-grown man, and you will see how age changes the aspect of the bones.

Fig. 150 is the hand of a full-grown woman. You will observe in the case of the lady's rings that the diamonds are transparent, while the metal portion casts a shadow even through the bones.

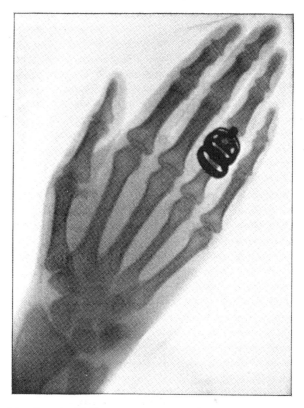

FIG. 150.—Hand of full-grown Woman. (*Photo. by Mr. J. W. Gifford.*)

These two photographs were taken by Mr. J. W. Gifford, of Chard, an early and most successful worker with Röntgen rays.

Fig. 151 is the hand of Lord Kelvin, and shows traces of age, and of a tendency to rheumatic deposits.

Fig. 152 is the hand of Mr. Crookes, and though a knottier hand, shows some points of resemblance with that of Lord Kelvin.

Fig. 153 is the hand of Sir Richard Webster. The shadow is interesting as showing not only an athletic development, but as revealing, embedded in the flesh between the thumb and the first finger, two small shot, the result of a gunshot wound received many years previously. This photograph and the two preceding are from the

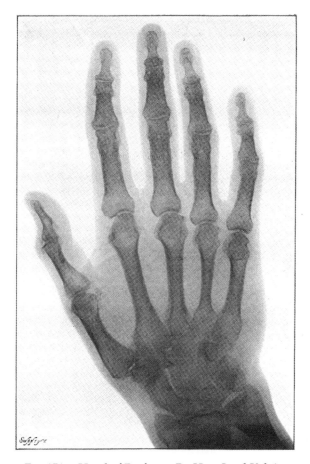

FIG. 151.—Hand of Professor Rt. Hon. Lord Kelvin.

series taken by Mr. Campbell-Swinton, who was first in England to put into practice this newest of the black arts.

By the courtesy of Mr. Campbell-Swinton I am also able to show you a number of other slides—the hand of Lord Rayleigh; the hand of a lady with a needle embedded in the palm; a hand terribly swollen with the gout; a foot, showing the heel-bone and the smaller bones down to the toes, as well as the bones in the ankle; a view through the left shoulder of a young lady, showing her ribs, shoulder-blade, and collar-bone; the torso of a young man, showing

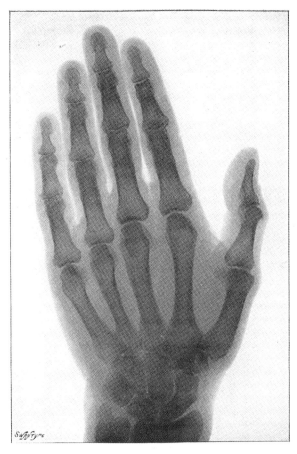

FIG. 152.—Hand of Professor W. Crookes, F.R.S.

his ribs, and, dimly, his heart, like a central dark shadow with a triangular apex pointing down toward the right, that is, to his left side;
lastly, the shadow of a living head, showing all the vertebræ of the
neck.

Here, again, is the shadow of a newly-born child, taken by Mr.
Sydney Rowland. Note the imperfect state of the bones in the hands.

Passing from human objects, we will look at the shadows of a
few animals. These are a chameleon, giving a clear view not only
of its skeleton but of the internal organs; a mouse; a frog; and some

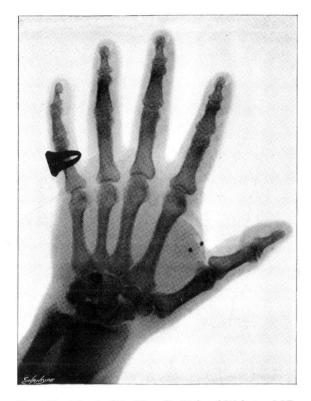

FIG. 153.—Hand of Rt. Hon. Sir Richard Webster, M.P.

fishes. The next slide was taken from an Egyptian mummy in its wrappings. Before this photograph was taken there was some dispute as to whether it was the mummy of a cat or of a girl. The photograph sets the question entirely at rest.

Earlier in my lecture I mentioned that glass is tolerably opaque to these rays. Of this you have a proof in the next photograph (Fig. 154), which represents a pair of spectacles photographed while lying in their case, the covering of which, in shagreen, shows all the markings peculiar to the shark's skin, with which the case was covered. The next photograph by Mr. Campbell-Swinton enables you to read the contents of a sealed letter which he received. His also is the next picture (Fig. 155), which is the photographed shadow of an aluminium

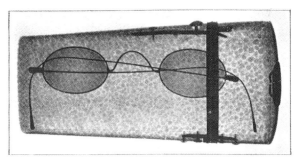

Fig. 154.

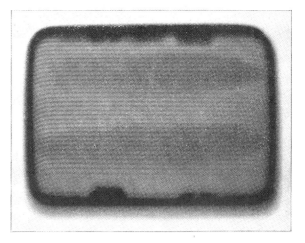

Fig. 155.

cigar-case, containing two cigars. And lastly (Fig. 156), I exhibit to you a photograph of two ruby rings. By gaslight the gems of one are not distinguishable from those of the other; and in broad daylight it would take an expert to pronounce between them. But when viewed or photographed by Röntgen light there remains no manner of doubt. The rubies of one ring are true Burmese rubies, and they appear transparent. The others are imitation rubies made of ruby-coloured glass, and appear quite opaque.

You will have noticed that I have spoken of these rays as "Röntgen light." But are we really justified in calling it light? It is

Fig. 156.

invisible to our eyes; but then so also is ordinary ultra-violet light, and so is infra-red light, and Hertzian light. And there are other kinds of light too, amongst them one discovered during last year by M. Becquerel[12] and myself, which are invisible. But if the Röntgen light can be neither reflected nor refracted, neither diffracted nor polarised, what reason have we for calling it light at all? In fact, direct proof that it consists of transverse waves is wanting. Many conjectures have been formed respecting its nature. Röntgen himself suggested that it might consist of longitudinal vibrations. Others have suggested ether streams, ether vortices, or even streams of minute corpuscles. At one time the notion that it might be simply an extreme kind of ultra-violet light of excessively minute wave-length was

---

[12] M. Henri Becquerel (see *Comptes Rendus*, cxxii. pp. 559, 790, etc.) and I myself (see *Philosophical Magazine*, July 1896, p. 103) quite independently discovered some invisible radiations that are emitted by uranium salts, and by the metal uranium, which can affect photographic plates, and will pass through a sheet of aluminium or of cardboard. But they are not the same as Röntgen rays, since, as M. Becquerel has shown, they can be reflected, refracted, and polarised. They also produce diselectrification. There can be no question that these rays, which are due to a sort of invisible phosphorescence, consist of transverse vibrations of a very high frequency: that is, they are ultra-violet light of a very high order.

favoured by physicists, who were disposed to explain the absence of refraction, and the high penetrative power of the rays upon von Helmholtz's theory of anomalous dispersion, according to which the ultra-violet spectrum at the extreme end ought to double back on itself.

The most probable suggestion yet made, and the only one that seems to account for the strange lateral emission of the rays right up to the plane of the antikathode (see Fig. 142, p. 265) [p. 148 here], is that of Sir George Stokes. Stokes's view is that while all ordinary light consists of trains[13] of waves (Fig. 68, p. 112) [not included here], in which each ripple is one of a series that gradually dies away, the Röntgen light consists of solitary ripples, each of not more than one or one and a half waves. According to Stokes the Röntgen light is generated at the antikathode by impact of the flying negatively-electrified molecules (or atoms) which constitute the kathode stream. At the moment when each of these flying molecules strikes against the target and rebounds, there will be a quiver of its electric charge; in other words, the charge on the molecule will perform an oscillation. Now that electric oscillation will be executed across the molecule in a direction generally normal to the plane of the target, and will give rise to an electro-magnetic disturbance which will be propagated as a wave in all directions, except where stopped by the metal of the target. And this oscillation being of excessively short period, and dying out after about one or two (Fig. 157) complete periods, will generate a wave, which, though of a frequency as high as, or even higher than, that of ordinary ultra-violet light, and therefore capable of producing

---

[13] It has long been known from the experiments of Fizeau, that in ordinary light each train consists on the average of at least 50,000 successive vibrations; for it is possible to produce interference of light between two parts of a beam which have traversed lengths differing by more than 50,000 wave-lengths. Michelson has gone far beyond that number. See the footnote on p. 112 [not included here].

FIG. 157.

FIG. 158.

kindred effects, will not be capable of being made to interfere, nor to undergo regular refraction or reflexion, because it does not consist of a complete train of waves. Here is a model intended roughly to illustrate the theory. An iron hoop (Fig. 158) which can be thrown or swung against the wall represents the flying molecule. The electric charge which it carries is typified in the model by a lump of lead capable of sliding on a transverse wire, and held centrally by a pair of spiral springs. When this model molecule is caused to strike against the wall and rebound, the leaden mass is disturbed, and executes an oscillation to and fro along the wire. The oscillation dies out after about $1\frac{1}{2}$ periods. Now, suppose this oscillation to set up a transverse wave in surrounding space. Though it consists of but $1\frac{1}{2}$ ripples, they would be propagated outward just as trains of waves are. And if there were millions of such flying molecules in operation, these solitary ripples might come in millions one after the other, but not regularly spaced out behind one another like the trains of waves constituting ordinary light. This is but a gross and rough illustration of Stokes's hypothesis; but it must suffice for the present.

But I cannot close this course of lectures without one word as to the possibilities which this amazing discovery of the Röntgen light has opened out to science. It is clear that there are more things in heaven and earth than are sometimes admitted to exist. There are sounds that our ears have never heard: there is light that our eyes will never see. And yet of these inaudible, invisible things discoveries are made from time to time by the patient labours of the pioneers in science. You have seen how no scientific discovery ever stands alone: it is based on those that went before. Behind Röntgen stands Lenard; behind Lenard, Crookes; behind Crookes the line of explorers from Boyle and Hauksbee and Otto von Guericke downwards. We have had Crookes's tubes in use since 1878, and therefore for nearly twenty years Röntgen's rays have been in existence, though no one, until Röntgen observed them on 8th November 1895, even suspected[14] their presence or surmised their qualities. And just as these rays remained for twenty years undiscovered, so even now there exist, beyond doubt, in the universe, other rays, other vibrations, of which we have as yet no cognisance. Yet, as year after year rolls by, one discovery leads to another. The seemingly useless or trivial observation made by one worker leads on to a useful observation by another; and so science advances, "creeping on from point to point." And so steadily year by year the sum total of our knowledge increases, and our ignorance is rolled a little further and further back; and where now there is darkness, there will be light.

---

[14] It is but fair to Professor Eilhard Wiedemann to mention that in August 1895 he described some "discharge-rays" (Entladungsstrahlen) inside a vacuum tube, which, though photographically active, refused to pass through fluor-spar, and were incapable of being deflected by a magnet. But their properties differ from Röntgen rays in some other respects.

The Great Extinct Reptiles—Dinosaurs from the Oolites—The Pariasaurus and Inostransevia from the Trias of North Russia and South Africa—Marine Reptiles

Edwin Ray Lankester*

In the next two chapters I propose briefly to bring before you a few examples of extinct reptiles, birds and fishes, and to take the very shortest glance at the host of invertebrate shell-fish, insects, star-fishes and such like extinct animals whose name is legion.

We will proceed at once to the reptiles. You will see from the list of groups of reptiles which I gave to you in a former chapter (p. 58) [not included here] that there are four big orders or groups of living reptiles: (1) the Crocodiles; (2) the Tortoises (Chelonians); (3) the Lizards; and (4) the Snakes. The lizards and snakes are in their real structure so much alike that they are considered as one double order. Extinct representatives of all these orders are found right away down through the Mesozoic strata to the Trias (see table of strata, p. 60) [not included here]. But there is nothing very astonishing about them excepting the large size of some of the extinct tortoises and snakes, and the fact that the older extinct crocodiles had the opening of the nose-passages into the mouth-openings, which we and all air-breathing vertebrates also possess, placed far forward as they are in the more primitive air-breathers, whereas living crocodiles have

* The original version appeared in: Edwin Ray Lankester, Chapter 5, *Extinct Animals*, London, 1905, pp. 190–244.

them pushed ever so far back to the very furthest recess of the long ferocious mouth, from which arrangement it results that the modern crocodile can have its mouth full holding the body of a victim under water whilst the air passes from the tip of its nose through the long nasal passage to the very back of its mouth and so to its lungs. This convenience was not enjoyed by primitive crocodiles.

The great interest in regard to extinct reptiles centres in those which were so entirely different from the reptiles of to-day that naturalists have to make separate orders for them. Many of them were of huge size. They flourished in the Mesozoic period and abruptly died out; at any rate their remains disappear from the rocks at the close of the Chalk or Cretaceous period (see the table of strata, p. 60) [not included here]. These extinct orders of reptiles are the Dinosaurs, the Theromorphs, the Ichthyosaurs, the Plesiosaurs and the Pterodactyles. They are a prominent example of that kind of extinct animal which is not the forefather, so to speak, of living animals, but of which the whole race, the whole order, has passed away, leaving no descendants either changed or unchanged.

To begin with the Dinosaurs. They are a very varied group and mostly were of great size. They seem to have occupied in many ways the same sort of place on the earth's surface which was filled at a later period by the great mammals, such as elephants, rhinoceroses, giraffes, giant kangaroos, etc. Preying on the vegetable-feeding kinds there were huge carnivorous dinosaurs, representing the lions and tigers of to-day. Yet the mammals I have mentioned are in no way descended from these great reptiles. They came from another stock, and only superseded them on the face of the earth by a slow process of development, in which the great reptiles disappeared and the great mammals gradually appeared and took their place.

Some of the forms assumed by the great Dinosaurian reptiles are not unlike the forms of the small scaly lizards of to-day (see Figs. 137, 138, 139, 140); but on the whole the Dinosaurs were more like mammals in shape, standing well up on the legs. We do not know much

FIG. 137.—Photograph of a cast taken from life of the New Zealand Lizard Tuatara, known as *Sphenodon punctatus*. The figure is one-third of the natural size.

FIG. 138.—*Phrynosoma orbiculare* (Mexican Horned Lizard or Horned Toad). Photographed one half the natural size.

about their skin; it was probably smooth and with only small horny scales on it, as in many living lizards, and often had great horns and crests growing out of it. But we know the complete skeletons put together from bones chiselled out of the hard rock in which they are found, and we know that in important matters of shape and build the skeleton was different from that of living reptiles. The great size

FIG. 139.—*Chlamydosaurus kingi*, from Queensland, Australia. Photographed to one-third the natural size.

FIG. 140.—*Zonurus giganteus* (Great Girdled Lizard). Photographed half the size of nature.

to which some of the Dinosauria attained is shown by the thigh-bone of one found in the United States, and called Atlantosaurus— photographed in Fig. 6, p. 11 [not included here]. This thigh-bone is one third as long again as that of the biggest elephant known.

In Fig. 141 is shown the complete skeleton of the Iguanodon. This great Dinosaur was one of the first to be discovered. As you see, it stood on its hind legs like a kangaroo, and in running occasionally went on those feet only, touching the ground now and then with its front feet. Footprints in slabs of sandstone, once soft wet sand, are found showing this. The animal stood about fourteen feet from the head to the ground in the position shown in the figure. Its thigh bone was only three feet long and it was therefore only half the size, in linear measurement, of the Atlantosaurus.

In Fig. 142 an attempt is made to show what the animal looked like when the skeleton was clothed with flesh and skin. The first bones and teeth of the Iguanodon were found seventy years ago by a celebrated and most delightful collector and explorer of the earth's crust, Dr. Gideon Mantell, in the strata known as the Wealden in Sussex, just below the Chalk and Greensand (see table of strata). Dr. Mantell found that the teeth, of which two are here represented of the natural size, were those of a herbivorous animal

FIG. 141.—Drawing of the skeleton of *Iguanodon bernisartensis*. From the ground to the top of the head as the animal is posed, is about fourteen feet.

Fɪɢ. 142.—Probable appearance of the Iguanodon in its living condition.

and like those of the little living lizard from South America, called the Iguana, in the fact that the broad chisel-like crown has a saw-like edge (Fig. 144). From this fact the name Iguanodon (Iguana-toothed) was given to the new fossil giant reptile. The bones found by Mantell and others were scattered and not in their natural position and the form of the creature had to be guessed at by fitting this and that together. But some twenty-five years ago a wonderful find was made near Brussels in a coal-mine at a village called Bernissart. The skeletons of no less than twenty-two huge Iguanodons were found complete, and embedded in a fairly soft clay-like rock! The authorities of the Government Museum took charge of the place and most care fully [*sic*] removed the rock containing the skeletons to the Museum workshops at Brussels, where

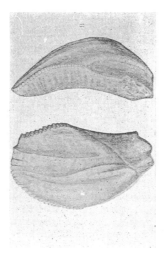

FIG. 143.—Two teeth of *Iguanodon mantelli* of the natural size, showing the serrated margin.

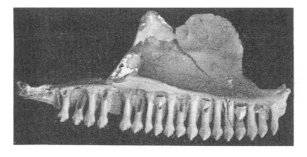

FIG. 144.—A portion of the upper jaw of the recent lizard Iguana, showing the serrated edges of the teeth, similar to those of Iguanodon.

the complete skeletons of seven were, with enormous difficulty and care, removed bit by bit from the rock and set up as entire skeletons in the Brussels Museum, where they may be seen. A cast of one of these seven is in our own Natural History Museum.

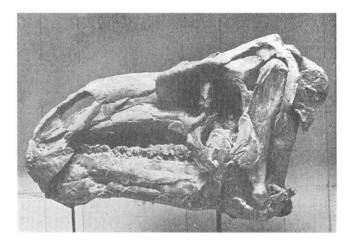

FIG. 145.—Photograph of the skull of an Iguanodon as dug out of the rock, showing the teeth of the lower jaw and the smooth bony supports for the horny beak of both upper and lower jaw. The specimen is three feet in length.

The photograph of the skull of one of these specimens is given in Fig. 145. It shows not only the teeth in position, but in front the bony supports of a great horny beak, like that of a turtle. As you may see in the drawing of the skeleton (Fig. 142), the forefeet (or hands) were provided with five fingers, of which the thumb had a huge claw on it at least a foot long. The foot was very much like that of a bird and had only three toes, and the bones of the pelvis or hip-girdle are extraordinarily like those of a bird. In fact it is now certain that reptiles similar to the Iguanodon were the stock from which birds have been derived, the front limb having become probably first a swimming flipper or paddle, and then later an organ for beating the air and raising the creature out of the water for a brief flight. From such a beginning came the feather-bearing wing of modern birds.

Fig. 146 shows the skeleton of a Dinosaur of somewhat less size but with the same kangaroo-like carriage, which was a beast of

FIG. 146.—Drawing of the skeleton of a carnivorous Dinosaur, the Megalosaurus. The animal was about two-thirds the size of the Iguanodon.

prey. It is the Megalosaurus, and had many tiger-like teeth in its jaws. It hunted down and fed upon the herbivorous Dinosaurs as lions and tigers hunt and eat antelopes and buffalo to-day. By no means all the Dinosaurs walked on their hind legs. There were enormous kinds which went on all fours. Here is the skeleton of the Brontosaurus (Fig. 147) and a sketch of its appearance in life (Fig. 148). The great Ceteosaurus, of which the limb bones and most of the skeleton were found near Oxford, is similar to this, and Mr. Andrew Carnegie has presented to the Natural History Museum a complete reconstruction of the skeleton of a closely allied Dinosaur—the Diplodocus—which was excavated in Wyoming and is now in the Carnegie Institute at Pittsburg. It is eighty feet long. Its head is very small, and a great part of the length is made up by the very long neck and the very long tail, but the body is bigger

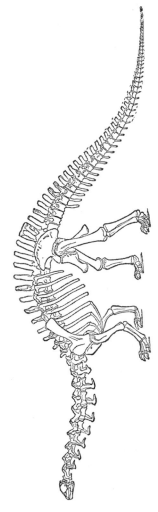

FIG. 147.—Drawing of a completely restored skeleton of the Brontosaurus. Note the extremely small size of the skull and the great length of the head and tail. A man could walk in front of the legs under the neck without stooping.

FIG. 148.—Probable appearance of the Ceteosaurus (and of the closely similar Diplodochus and Brontosaurus) in life. It has been suggested that the animal walked along the sea or river bottom keeping its head just above water. Specimens of over sixty feet in length have been found.

than that of the biggest elephant and the back was nearly fourteen feet from the ground.

The immense profusion in which the bones of these huge creatures have been found in Mesozoic strata in the United States is astonishing; no less remarkable is the skill and success with which American naturalists—chief among whom have been Professor Marsh of Yale and Professor Cope of Philadelphia—have collected, fitted together and drawn every detail of more than thirty different kinds of these monsters. They have given such full evidence of the structure and build of the animals that we may with confidence accept the reconstructions of the appearance of the animals such as those shown in Figs. 149 and 150, where the rhinoceros-like Triceratops and the huge crested Stegosaurus are represented. Such crests and horns are bizarre and grotesque even when carried by little living lizards a few inches long, but it must be remembered

FIG. 149.—Drawing of the appearance in life of the three-horned Dinosaur, Triceratops (after a model issued by the American Museum of Natural History). This reptile was of the size of the largest living Rhinoceros.

that the Dinosaurs drawn in Figs. 149 and 150 were as big in the body as large elephants.

A curious fact about these great Dinosaurs is that they had, as compared with big living reptiles such as the crocodiles, very tiny brains. You will remember that the extinct mammals known as Titanotherium and Dinoceras have brains one-eighth the bulk of living mammals of the same size, such as rhinoceros and hippopotamus. So it was with the huge extinct reptiles. In some the head itself was ridiculously small according to our notions of customary proportion, and even in others, such as Triceratops, where the bony and muscular parts of the head were big, as in a rhinoceros, yet the brain was incredibly small. It could have been passed all along the spinal canal in which the spinal cord lies, and was in

FIG. 150.—Probable appearance in life of the Jurassic Dinosaur Stegosaurus. The hind leg alone is twice as tall as a well-grown man.

proportion to bulk of body a tenth the size of that of a crocodile. Very probably this small size of the brain of great extinct animals has to do with the fact of their ceasing to exist. Animals with bigger and ever increasing brains outdid them in the struggle for existence.

So much for the Dinosaurs, which might well occupy a complete course of lectures all to themselves. We will now turn to the Theromorphs, which are an older group even than the Dinosaurs and flourished in the Trias period (see table of strata, p. 60) [not included here]. The Theromorphs are so called because in some important parts of the structure of skull and jaw, and often also in the teeth, they resemble the mammals or Theria. They come near to a point in the history of terrestrial vertebrate beasts which is the common origin of Reptiles, Mammals and Batrachia or Amphibians (newts, salamanders and frogs).

Their remains have been found in the Triassic sandstones and limestones of South Africa, of Russia, of India and of Scotland and

Fig. 151.—Photograph of the skeleton of Pariasaurus, as set up by Professor Seely in the Natural History Museum. Length from snout to tail, about eight feet.

the centre of England. One of the most striking of these is represented by a completely reconstructed skeleton from Cape Colony in the Natural History Museum, photographed in Fig. 151. The skeleton is some eight feet long and looks like a gigantic pug-dog. This is the Pariasaurus, and is shown by its small teeth to have been herbivorous.

From the same locality we have the Dicynodon with two huge tusks, and the Cynognathus with a skull and set of teeth wonderfully recalling those of a bear at first sight.

Another strange crested form belonging here is the Dimetrodon from the Permian strata of Texas, U.S.A. (Fig. 152).

But I am now able to show you, through the kindness of Professor Amalitzky, of Warsaw, a set of photographs taken by him, showing the discovery and working out by him of a whole series of skeletons of these Theromorph reptiles, closely similar to those from the rocks of Cape Colony but belonging to a locality far

FIG. 152.—Probable appearance in life of the Theromorph Reptile, Dimetrodon, from the Permian of Texas. As big as a large dog.

removed from South Africa, namely, to the banks of the Northern Dwina near Archangel in North Russia. Professor Amalitzky has not yet finished his excavations nor published these photographs, and it is therefore a great kindness on his part to allow me to show them here in London.

First of all, we have the cliff of Permian strata on the banks of the Dwina (Fig. 153), from which and from another similar spot the remains were extracted. At this point, where the colour is dark in the photograph, there is a peculiar "pocket" or accumulation of sandy matter with large hard nodules embedded in it. These nodules are removed and broken up for mending the roads. The pocket seems to be in a fissure and of Triassic age, later, that is to say, than the Permian rocks on each side of it. However that may be, the great nodules are removed from it for road mending, and four or five years ago Professor Amalitzky on visiting the spot was astounded and delighted to find that when broken each nodule

FIG. 153.—View of one of the dark patches in the cliffs of the river Dwina (the Northern of that name), where nodules containing the skeletons of extinct reptiles are found.

FIG. 154.—One of the nodules showing the form of the embedded skeleton, head to the right, tail to the left.

was seen to contain the skeleton or skull of a great reptile. Fig. 154 shows such a nodule, some eight feet long, and in this specimen one can easily distinguish the skull, the four limbs and the back-bone of a large animal. The Russian geologist determined to make a most thorough investigation of this wonderful deposit, and for

FIG. 155.—Peasants working on the face of the cliff near Archangel and removing nodules containing the skeletons of great reptiles.

some years now has spent a thousand pounds a year, obtained for the purpose through the Imperial Academy of St. Petersburg, in having the nodules dug out by the peasants after their farming work is over for the year, and in removing them to the University of Warsaw, where with the finest instruments and greatest care the nodules are opened and each bone removed in fragments is put together from its more or less broken parts, firmly cemented and set up in its natural position and relations as part of a complete skeleton. Fig. 155 shows the peasants at work, protected by a shed from the fall of stones from above. Fig. 156 shows some of the nodules as yet unopened lying in the laboratory of the geological professor at Warsaw. Fig. 157 shows a number of skeletons of the huge but harmless vegetarian Pariasaurus which have been cleared out of the nodules and set up on iron supports, as more or less complete specimens. Of course it is not possible in every individual to

FIG. 156.—Professor Amalitzky's work-shop in Warsaw, showing skeleton-holding nodules ready to be broken open and others already under preparation.

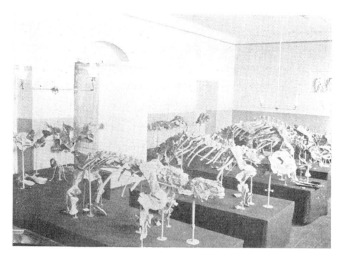

FIG. 157.—A series of skeletons of Pariasaurus removed bit by bit from Archangel nodules and mounted as detached specimens by Professor Amalitzky.

get out all the bones complete, especially those of the feet. Few of the individuals were complete even when originally embedded in the mud ages ago. When an animal's body is carried away by a river and floats in a decomposing state it tends to fall to pieces.

The cliff formed by the present river Dwina consists of rocks of immense, indeed of almost inconceivable, age, and existed as solid rock ages and ages before the surface of the earth had its present form. These deep-lying rocks have been brought near to the surface by bending of the strata (as shown in Fig. 36, p. 52) [not included here], and the cutting or cliff made by the comparatively modern river exposes them to our view and to easy excavation. The nodules are relatively to the age of the river-valley or cutting (which is probably some 150,000 thousand years old), as much older than it is as are Roman coins older than the trench dug three hours ago which brings them to light. If you look at the position of the Trias and Permian in the table of strata you will get some idea of how immensely remote is the time when these great reptiles lived where now is Archangel, for whilst the thickness of a twentieth of an inch suffices to indicate the accumulations of strata since the mammoth lived in England, the Trias is a long way down the series, far below the Eocene, where the ancestral elephants of Egypt are found, far below the Chalk, and older than the long Jurassic series of rocks in which the remains of the great Dinosaurs we have recently looked at, occur.

In Fig. 158 one of Professor Amalitzky's specimens of Pariasaurus is shown. There is no artificial completing of this skeleton: all that is seen is actual bone as cleaned out of a nodule. Only one foot is preserved, but that of course tells us as to its fellow of the opposite side. The skull of another specimen of Pariasaurus is shown in Fig. 159. It is very remarkable that this species seems to be so closely similar to the one discovered far away in South Africa in beds of the same age.

These Pariasaurs were about as big as well grown cattle, but not so high on the legs. In Fig. 160 we have the skeleton of another

FIG. 158.—Photograph of a skeleton of Pariasaurus, removed from an enveloping nodule and mounted by Professor Amalitzky.

FIG. 159.—Photograph by Professor Amalitzky on a larger scale of a skull of a Pariasaurus from an Archangel nodule.

creature revealed by these nodules. It is an enormous and truly terrible carnivor, with a skull two feet long and enormous tiger-like teeth. This creature is named Inostransevia by Professor Amalitzky, and is larger than any of the carnivorous reptiles from South

FIG. 160.—Skeleton of a huge carnivorous beast of prey — the reptile named Inostransevia, discovered and photographed by Professor Amalitzky of Warsaw. The skull alone is two feet in length.

FIG. 161.—Skull of the gigantic Theromorph Carnivorous Reptile, Inostransevia discovered by Professor Amalitzky in Northern Russia. It is allied to Lycosaurus found in Cape Colony in beds of the same age.

Africa. Specimens of its skull are shown in the Professor's photographs reproduced in Figs. 161 and 162. No doubt the vegetarian herds of Pariasaurus, whose small peg-like teeth indicate clearly enough their inoffensive habits, were preyed upon by the terrible

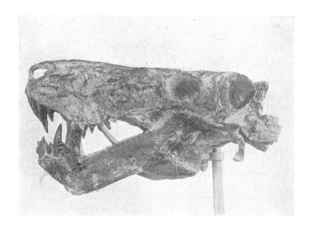

FIG. 162.—Photograph of another skull of Inostransevia.

Inostransevia, as were their brethren in South Africa devoured by the Cynognathus, the Lycosaurus, the Cynodraco and other carnivorous reptiles of that remote Triassic age. So we see the co-existence of blood-sucker and victim—of the destructive oppressor and the helpless oppressed—forced on our attention in these two localities, Russia and South Africa, when we study the immensely remote past of the Triassic age.

We leave now these great extinct land-dwelling reptiles and take a glance at representatives of two extinct orders of huge aquatic creatures which must also be classified as reptiles. These are the Plesiosauria and the Ichthyosauria. Though some of them must have measured thirty feet from snout to tail, they do not equal in size the great aquatic mammals of to-day, the whales.

In Fig. 163 is shown the photograph of the skeleton of a large Plesiosaur, and in Fig. 164 is given a drawing showing how the creature appeared in life. It had a body like the hull of a submarine with four paddles attached, the fore- and the hind-legs. It had a long neck like that of a swan and an elongated head provided with powerful jaws armed with numerous pointed teeth. It probably

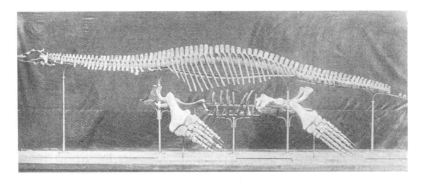

FIG. 163.—Photograph of a skeleton of a Plesiosaur removed by Mr. Alfred N. Leeds from the clay (of Jurassic age) near Peterborough, and set up bone by bone in a nearly complete state in the Natural History Museum.

FIG. 164.—Plesiosaurus as it probably appeared when alive, swimming near the surface of the water with its back showing and its neck and head raised above the surface.

could swim under water as well as on the surface, and when in the latter position could snap small lizards and birds from the land. The paddles have the definite structure of legs, with five toes, wrist or ankle and fore-arm or fore-leg and upper arm or thigh. A great

FIG. 165.—Photograph of a skeleton of the large-paddled Ichthyosaurus preserved in Liassic rock.

number of kinds of these Plesiosaurs have been discovered, especially in the Lias rocks of the South of England, slabs containing whole skeletons being frequently obtained. They and the similarly embedded and flattened skeletons of different kinds of Ichthyosauria may be seen in quantity on the walls of the gallery of fossil reptiles in the Natural History Museum.

In Fig. 165 the flattened skeleton of an Ichthyosaurus is photographed. This particular species is remarkable for the great size of its fore-paddles.

In Fig. 166 a drawing of an Ichthyosaurus, as it must have appeared in life, is given. The Ichthyosaurs are much more fish-like or rather whale-like in form than the Plesiosaurs. They were indeed singularly like the porpoises and grampuses among living whales and stand in the same relation to land-living reptiles that the porpoises do to land-living mammals. Their fish-like appearance and fins are not primitive characters and do not indicate any closer blood-relationship to fishes than that possessed by other reptiles. They are the offspring of four-legged terrestrial reptiles which have become specially modified and adapted to submarine life. Like many whales they had a median fin on the back devoid of bony support. The bones of their legs have become greatly

FIG. 166.—Drawing to show the probable appearance of an Ichthyosaurus swimming beneath the surface of the sea.

changed, much more so than those of the Plesiosaurs and form often more than five rows of nearly circular or polygonal plates fitted together as a flexible paddle. The tail is fish-like, but has the lower lobe bigger than the upper and the vertebral column bends down into the *lower* lobe instead of turning up into the *upper* lobe as it does in fish. The details as to the fins are known from some wonderfully preserved specimens found in the fine hardened mud known as the lithographic slate of Solenhofen, where the soft bodies of jelly-fish, cuttle-fishes and the wings of flying reptiles also are preserved.

As mentioned in the first chapter, the Ichthyosauria (see Fig. 2) [not included here] had a ring of bony plates supporting the eyeball (as birds also have), and these are often preserved in the fossil specimens. In Fig. 168 [*sic*; Fig. 167] a view of the top of the skull of an Ichthyosaurus is given in order to show the round hole in the middle line of the brain-case (on a level with the letter P). This is

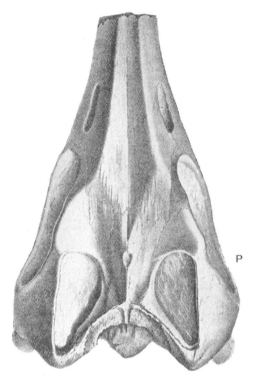

FIG. 167.—Photograph of the upper surface of the skull of an Ichthyosaurus. On a level with the letter P in the middle of the skull is seen an oval pit, the "parietal foramen" in which was lodged the "third" or "pineal" eye.

called the "parietal foramen," and is a fair-sized hole in which was lodged an eye, a third eye called the pineal eye. This eye is found in some other reptiles also, and especially in some of the living lizards where its structure has been studied with the microscope. There is no doubt that the body filling this hole in living lizards is an eye, although it seems to have lost the power of sight in these recent forms. A third eye, placed on the top of the head strikes one as a very strange arrangement and contrary to all our common experience of vertebrate animals.

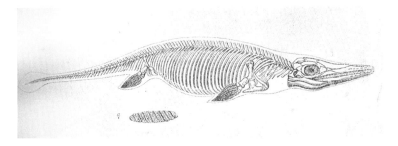

Fɪɢ. 168.—Side view of the skeleton of an Ichthyosaurus. Below the skeleton is drawn a "coprolite" showing spiral grooving on its surface.

In Fig. 168 we have a drawing of the side view of the skeleton of an Ichthyosaurus and below it a fossilized lump of its excrement. These are called coprolites and consist of scales and bones of fishes digested by the Ichthyosaurus. They show a corkscrew-like moulding of the surface, proving that the intestine of the Ichthyosaurus had a spiral fold like a spiral staircase on the walls of the intestine, as have the sharks. We also find within well preserved specimens of the skeletons of Ichthyosaurus the skeletons of unborn young individuals, showing that the Ichthyosaurus brought forth its young alive.

We pass on now to even more astonishing reptiles—the extinct order of the pterodactyles or flying reptiles of the Mesozoic period. These creatures were as truly aërial as the birds and bats of to-day. They were of many kinds, from the size of a crow to so huge a form as that drawn in Fig. 169, which measured eighteen feet from the tip of one wing to the tip of the other. Their wings have been found well preserved in the Lithographic slates (see Fig. 31, p. 47) [not included here], and each consisted of a membrane spread from one enormously big elongated finger to the side of the body and little hind legs.

Fig. 170 gives some idea of the form and appearance of the wings when expanded. Such a wing is more like that of a bat than that of a bird, since it is a membranous skin and not a series of feathers.

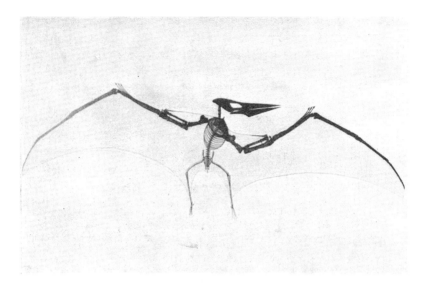

FIG. 169.—Photograph of a restoration of the skeleton of the great Pterodactyle (*Pteranodon*). The stretch of the wings from tip to tip measures eighteen feet. From a preparation in the Natural History Museum. The bones of the arms, shoulder girdle and fingers are the actual bones; the skull, neck, body and hind legs being drawn from other specimens of the same species.

FIG. 170.—The great Pterodactyle *Pteranodon* as it appeared in flight.

A.

B.

C.

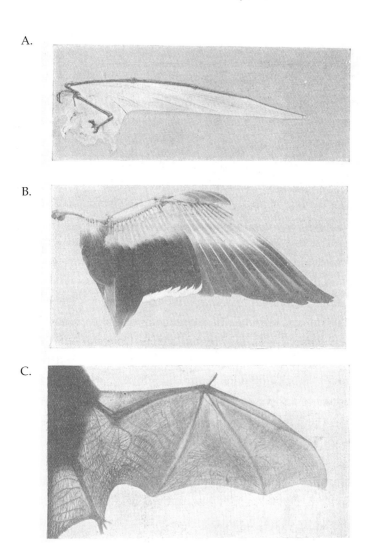

FIG. 171.—Photographs of three wings for comparison of their structure. A. That of a Pterodactyle, membrane supported by one long finger. B. That of a Bird, feathers set on the fore-arm (*cubitus*) and hand. C. That of a Bat, membrane supported by three elongated fingers.

The bat's wing is a membrane supported by three of the fingers as well as the side of the body and hind leg.

In Fig. 171 the fossil wing of a Pterodactyle, that of a recent bird with the bones and the great quill-feathers only in place (the smaller feathers having been plucked off), and the wing of a bat are photographed and placed together for comparison. There are two other kinds of flying animals, namely, the flying fishes (which do not fly far), and the six-legged insects or flies, bees and beetles. They have all independently acquired the habit of flying and have had certain parts of their bodies changed into wings. The process of change must have been gradual and have taken an enormous lapse of time to bring it about in each kind. There are some animals, such as the flying squirrels and flying lizards (*Draco volans*) of to-day, which do not really fly, since they have no wings to beat the air with, but can spread out a great flat surface on each side of the body which enables them to sail through the air for some distance without falling when they jump from the branch of a tree. This, however, is a long way from the point reached by animals which have wings and can strike the air as a fish strikes the water with its fins. Probably the wings of birds and of insects were both derived from fin-like organs which were used to swim with— before they were used in the air. But the origin of the wing of the Pterodactyles, and independently that of the wing of the bats, does not seem to have been of this nature, and is one of the many very puzzling matters which further discoveries may one day enable us to understand.

In Fig. 172 two other kinds of Pterodactyle are shown. Some Pterodactyles had no teeth, but long beak-like jaws (Fig. 169). Others had numerous sharp-pointed teeth and were beasts of prey.

It seems natural to pass from the winged reptiles to birds. But as a matter of fact the birds are not very closely related to Pterodactyles. Birds are, it seems, derived from reptiles, and are very specialized, warm-blooded descendants of certain reptiles.

FIG. 172.—Probable appearance in life of two kinds of Jurassic Pterodactyles (*Dimorphodon* and *Rhamphorhynchus*).

They are so peculiar that they are considered as a distinct "class," and the reptiles which come nearest to them in structure are the Dinosaurs, especially those Dinosaurs (like Iguanodon) which walked on their hind-legs and had only three toes to the foot. Fossil remains of birds are not abundant—but a few very interesting birds have been found in the Lower Eocene and in the Cretaceous rocks (see list of strata, p. 60) [not included here], and one more remarkable than any other in the Lithographic slates of Jurassic age. Modern birds have all got feathers and beaks, and, with one or two rare exceptions, the quill feathers are set on the fore-arm and hand so as to form the wing. No living bird has teeth, but fossil birds are known with well developed teeth like those of reptiles. In Fig. 173 is shown the drawing of the skeleton of an extinct bird, which had a full set of teeth. The most remarkable extinct bird as

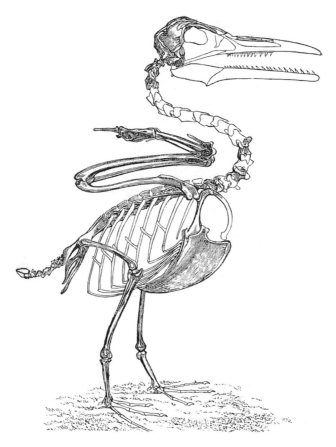

FIG. 173.—Restored skeleton of the toothed Bird, *Ichthyornis*—of the size of a pigeon—from the Chalk of Kansas, U.S.A.

yet discovered is that shown in Fig. 174. Two specimens of it have been obtained from the Lithographic slates of Solenhofen in Bavaria. The first one found is preserved in the Natural History Museum; the second and more perfect is in Berlin. This bird—called Archaeopteryx—was of the size of a large pigeon, had a short head apparently without a beak, and its jaws were armed with teeth. Whereas living birds have the fingers of the hand aborted and tied together, this bird had three distinct fingers, each

Fig. 174.—The Berlin specimen of the *Archoeopteryx siemensi*, showing the wings with three fingers, the long tail, the head and neck and the feathers of the wings and tail.

armed with a claw. Its legs were like those of living birds, and it had four toes. Its tail was unlike that of any living bird, and like that of a lizard. Whereas the bony part of the tail of living birds is very short and bears the tail feathers set across it fan-wise, the Archaeopteryx had a long bony tail made up of many vertebrae, and the feathers were set in a series one behind the other on each side of it, so that the tail resembled the leaf of a date palm in shape. Strange as this little creature appears, it was a genuine bird, for it had true feathers well developed, which are clearly shown in the two fossil specimens. Besides the two rows of feathers on the long tail, there are the full set of feathers spreading from the fore-arms and hands to form the wings, and the thighs also were covered with feathers.

Fɪɢ. 175.—Photographs to one scale of the South American Cariama and the skull of the gigantic extinct Phororachus.

It cannot be said that this ancient extinct bird goes far towards connecting birds with reptiles: but in the possession of separate claw-bearing fingers, a long bony tail and teeth, in the apparent want of a beak, it does come nearer to lizard-like reptiles than does any other known bird.

In the Tertiary Strata remains of various birds have been found. One of great interest on account of its enormous size is the Phororachus of South America. We have in Fig. 175 a photograph of the skull of this bird placed beside the stuffed skin of a living South American bird, the Cariama or Screamer. If the extinct bird had the general proportions and habits of the Cariama, as seems probable, it must have been a terrible monster, standing some twelve feet high and far exceeding the most powerful eagles and vultures in strength and the size of its beak and claws. Great extinct wingless birds are found in quite recent "alluvial" deposits in New Zealand and in Madagascar. The discovery of the bones of the great Moa of New Zealand has already been mentioned in our second chapter (p. 69) [not included here]. Many species of Moa have been found in New Zealand. The Moa is allied to the ostriches of Africa, the emeus and cassowaries of Australia, and the rheas of South America.

It appears that under certain conditions of life birds may gradually lose the use of their wings, which dwindle in size and finally may disappear altogether. Such wingless birds are not necessarily of one stock. The wingless condition, or the great reduction in the size of the wings, has occurred in various kinds of birds at various periods of the earth's history, and in the same way wingless insects of different orders have come into existence. In New Zealand, besides the Moas, which are all now extinct, a small kind of wingless bird is found which is still alive and is known as the Apteryx

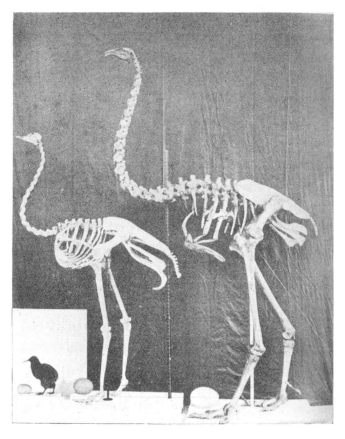

FIG. 176.—Photographs to one scale of the Apteryx, the Ostrich and the giant Moa of New Zealand, each with its egg.

or Kiwi. In Fig. 176 we have placed one behind the other each with its egg in front of it: a Kiwi, the skeleton of a very fine Ostrich, and the skeleton of a giant Moa (*Dinornis maximus*). The Polynesian islanders who landed in New Zealand some five hundred years ago, found the Moas still living, and hunted them down and lived upon their flesh. Skin and feathers of these enormous birds have been found preserved in a dried condition as well as the skeletons, and there are traditions as to the hunting of the Moa still in existence. The Moa of Madagascar seems to have been a smaller bird, but laid a proportionately much larger egg. It will be seen in Fig. 176 that the eggs of the Ostrich and of the Dinornis are not nearly so big in proportion to the size of the bird as is that of the Apteryx, which lays a truly gigantic egg considering the size of its body. The Moa of Madagascar is known as the Æpyornis and laid the biggest egg known—much bigger than that of the biggest New Zealand Moa—resembling the Apteryx in the proportionate sizes of its egg and its body. It was this very large egg which inflamed the imagination of ancient navigators and led to the vast exaggeration, which thrills the reader with wonder and terror, in the accounts of the "roc" given by Sinbad the Sailor in the *Arabian Nights*.

Flightless birds necessarily, unless they are, like the penguins, great swimmers, must get destroyed and become extinct when man arrives on the scene. The dodo, of which I spoke in my first lecture (p. 26) [not included here], was a close ally of the pigeons, but had lost its power of flight owing to the fact that it had no dangerous enemies in the island of Mauritius. It had become a heavy, slow running, though powerful ground bird. As soon as man arrived, and with him the pig, the flightless dodo was doomed to extinction. An extinct water-bird, the Hesperornis, had no wings whatever, whilst the penguins use their wings as swimming organs and are unable to fly. This was also the case with the garefowl or great auk (Fig. 15, p. 23) [not included here], which has recently become extinct.

The Atoms of Which Things Are Made

William Henry Bragg*

NEARLY two thousand years ago, Lucretius, the famous Latin poet, wrote his treatise *De rerum natura*—concerning the nature of things. He maintained the view that air and earth and water and everything else were composed of innumerable small bodies or corpuscles, individually too small to be seen, and all in rapid motion. He tried to show that these suppositions were enough to explain the properties of material things. He was not himself the originator of all the ideas which he set forth in his poem; he was the writer who would explain the views which were held by a certain school, and which he himself believed to be true. There was a rival set of views, according to which, however closely things were looked into, there would be no evidence of structure: however the water in a bowl, let us say, was subdivided into drops and then again into smaller drops and so on and on, the minutest portion would still be like the original bowl of water in all its properties. On the view of Lucretius, if subdivision were carried out sufficiently, one would come at last to the individual corpuscles or *atoms:* the word atom being taken in its original sense, something which *cannot be cut.*

* The original version appeared in: William Henry Bragg, Lecture 1, *Concerning the Nature of Things: Six Lectures Delivered at the Royal Institution*, London, 1925, pp. 1–41.

There is a mighty difference between the two views. On the one, there is nothing to be gained by looking into the structure of substances more closely, for however far we go we come to nothing new. On the other view, the nature of things as we know them will depend on the properties of these atoms of which they are composed, and it will be very interesting and important to find out, if we can, what the atoms are like. The latter view turns out to be far nearer the truth than the former; and for that all may be grateful who love to enquire into the ways of Nature.

Lucretius had no conception, however, of atomic theories as they stand now. He did not realise that the atoms can be divided into so many different kinds, and that all the atoms of one kind are alike. That idea is comparatively new: it was explained with great clearness by John Dalton at the beginning of the nineteenth century. It has rendered possible the great advances that chemistry has made in modern times and all the other sciences which depend on chemistry in any degree. It is easy to see why the newer idea has made everything so much simpler. It is because we have to deal with a limited number of sorts only, not with a vast number of different individuals. We should be in despair if we were compelled to study a multitude of different atoms in the composition of a piece of copper, let us say; but when we discover that there is only one kind of atom in a piece of pure copper, and in the whole world not many different kinds, we may feel full of enthusiasm and hope in pressing forward to the study of their properties, and of the laws of their combinations. For, of course, it is in their combinations that their importance lies. The atoms may be compared to the letters of the alphabet, which can be put together into innumerable ways to form words. So the atoms are combined in equal variety to form what are called molecules. We may even push the analogy a little further and say that the association of words into sentences and passages conveying meanings of every kind is like the combination of molecules of all kinds and in all proportions to form structures and materials that have an infinite variety of appearances and properties and can carry what we speak of as life.

The atomic theory of Lucretius did not contain, therefore, the essential idea which was necessary for further growth and progress. It withered away, and the very atom came to be used in a vague incorrect fashion as meaning merely something very small: as sometimes in Shakespeare's plays, for instance. In another and very different application of "atomic" theory Lucretius was strangely successful. He had the idea that disease was disseminated by minute particles. At the time of the Renaissance Fracastoro was inspired by the atomic theory of infection as he read it in the poem of Lucretius; but after his day the secret of bacteriology was again covered up until it was laid bare by Pasteur.[1]

Let us think of Nature as a builder, making all that we see out of atoms of a limited number of kinds; just as the builder of a house constructs it out of so many different kinds of things— bricks, slates, planks, panes of glass, and so on. There are only about ninety sorts of atoms, and of these a considerable number are only used occasionally. It is very wonderful that all the things in the world and in the universe, as far as we know it, are made of so few elements. The universe is so rich in its variety, the earth and all that rests on it and grows on it, the waters of the seas, the air and the clouds, all living things that move in earth or sea or air, our bodies and every different part of our bodies, the sun and moon and the stars, every single thing is made up of these few kinds of atoms. Yes, one might say, that is so: but if the builder is given bricks and mortar and iron girders he will build you an

---

[1] See "The Legacy of Rome" (Oxford University Press), p. 270 — an article by Dr. Singer.

infinite variety of buildings, palaces or cottages or bridges; why may not Nature do something like that? But one has to think that when a builder sets out to make a structure he has a plan which has cost thought to devise, and he gives instructions to his work-men who are to carry out his wishes, and so the structure grows. We see him walking about with his plans in his hand. But the plans of the structures of Nature are locked up in the atoms them-selves. They are full of wonder and mystery, because from them alone and from what they contain grows the infinite variety of the world. How they came to be such treasure-houses we are not ask-ing now. We ask ourselves what these atoms are like: we have been asking the question ever since their exceeding importance began to be realised more than a hundred years ago. Have they size and form and other characteristics such as are possessed by bodies with which we are familiar? We must look into these points.

But first let us realise that in the last twenty-five years or so we have been given, so to speak, new eyes. The discoveries of radioac-tivity and of X-rays have changed the whole situation: which is indeed the reason for the choice of the subject of these lectures. We can now understand so many things that were dim before; and we see a wonderful new world opening out before us, waiting to be explored. I do not think it is very difficult to reach it or to walk about in it. In fact, the new knowledge, like all sudden revelations of the truth, lights up the ground over which we have been travel-ling and makes things easy that were difficult before. It is true that the new lines of advance now open lead the way to fresh difficul-ties: but therein lies the whole interest and spirit of research. We will try to take the first steps into the new country so that we may share in the knowledge that has already come, and comes in faster every day.

We go back to our questions about the atoms. Before the new period set in remarkably accurate answers had already been given to some of them, at least. In this theatre of the Royal Institution, Lord Kelvin gave several addresses which dealt with the properties of atoms, and especially with their sizes. By several most ingenious and indirect devices he arrived at conclusions which we are now able to test by accurate methods; and we find that he was remarkably close to the truth. It was, of course, far more difficult to say what was the size of any particular atom than it was to say how much larger one atom was than another. For instance, the sizes of the atoms of potassium and carbon could be roughly compared by taking into account the relative weights of equal volumes of the solid potassium metal and of diamond which is a form of pure carbon. Potassium is lighter than water, the diamond is three and a half times as heavy. We know from chemical observations that the individual potassium atom is rather more than three times as heavy as the carbon atom. If we suppose that the packing of the atoms in the two cases is the same (as a matter of fact, we now know that it is only approximately so) we must conclude that the atoms in the metal potassium are much larger than the carbon atoms in the diamond, because, though heavier individually, they pack so as to make a lighter material.

To make a reasonable estimate of the actual size of any one atom is a much more difficult matter, but all the four lines of reasoning which Kelvin employed led him to very nearly the same result. "The atoms or molecules of ordinary matter must be something like the 1/10,000,000th or from the 1/10,000,000th to the 1/100,000,000th of a centimetre in diameter."[2] Our new methods tell us that the diameter of the carbon atom in diamond is 1.54 hundred millionths of a centimetre and that of the atom in the metal potassium is 4.50 hundred millionths. We see that Lord Kelvin's

---

[2] From a Friday Evening Discourse before the Royal Institution of Great Britain, March 4th, 1881.

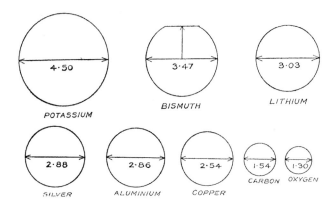

FIG. 1.—Sections of some common atoms, in hundred-millionths of a centimetre. In reference to bismuth see below (p. 12) [p. 206 here] and Plate I A.

estimate was wonderfully near the truth, considering the indirect and inexact methods which alone were at his disposal.

In Fig. 1 are shown sections of certain atoms on a scale of fifty millions to one. The inserted figures give in each case the distance, in hundred-millionths of a centimetre, between the centres of two neighbouring atoms in the pure substance. For example, the distance between two carbon atoms in the diamond is 1.54 hundred-millionths of a centimetre. In the case of oxygen the diameter has been calculated from the structure of crystals in which oxygen occurs. If the lecture-room of the Royal Institution were magnified as much as the atoms of Fig. 1, its height would be greater than the distance from the earth to the moon. We need some such comparison as this to make us realise the excessive smallness of the things of which we are talking. At the same time, we must keep in mind that they are not negligible because they are small: they are the actual elements of construction of the world and of the universe, and their size has nothing to do with their importance. But their smallness accounts readily for the ease with which we all overlook them, and for the difficulty we have in examining them when at last we have realised what they mean to us. The value of the new methods of which I propose to speak lies in the fact that they enable us to deal with them although they are so small.

We have now answered in a way the question as to the size of the atoms; but when we go further and ask ourselves about the shape we are not so successful.

The chemist, whose science is immediately concerned with the combinations of atoms, has rarely found it necessary to discuss their shapes, and gives them no particular forms in his diagrams. That does not mean that the shapes are unimportant, but rather that the older methods could not define them. There is one sense, however, in which the chemist pays much attention to form. The atoms in a compound are arranged in some fashion or other which is important to the combination. If one could see it and sketch it, one would be obliged to show it in perspective. In the science of organic chemistry especially it is found to be necessary to imagine such arrangements in space. It is not enough to represent them on the flat with no perspective at all; in fact, it is obvious that any flat design must be imperfect in any sort of chemical picture. We are unfortunately compelled to use the flat for our drawings; solid models in space are costly to make, while paper and pencil are cheap. It is curious to reflect what a handicap this technical difficulty puts on the proper development of a very important matter. Now, when we come to prescribe the arrangements of the atom to its neighbours, and to say that if one neighbour lies in this direction, another must lie in that, we are, in effect, giving shape to our atoms; at any rate, it is all the shaping that can be done for the present. We cannot do more until we know more about the internal structure of the atom: what its parts are, and how they are disposed to one another.

In the newer work, as we shall see, the arrangement of the atoms is much more closely examined, and for the first time their actual distances apart are measured. We find it absolutely necessary to make models because we do not see with sufficient clearness if we

are content to draw on paper. We represent our atoms as round
balls, and we find that we are able to represent most of our discoveries in this way. This really means that when an atom has several neighbours of the same kind it is equally distant from them all;
and this is actually the case. Nevertheless, there are exceptions, as
in the crystal of pure bismuth, where each atom has six neighbours
and three of them are closer than the other three. We have to make
a ball with three flats on it for use in constructing the bismuth
model (Plate I A).

Let us now ask ourselves what binds the atoms together into the
various combinations and structures. Like our builder, we have got
in our materials—the bricks, slates, beams and so on; we have our
various kinds of atoms. If we look round for mortar and nails we
find we have none. Nature does not allow the use of any new material as a cement. The atoms cling together of themselves. The
chemist tells us that they must be presented to one another under
proper conditions, some of which are very odd; but the combination does take place, and there is something in the atoms themselves which maintains it when the conditions are satisfied. The
whole of chemistry is concerned with the nature of these conditions and their results.

The atoms seem to cling to one another in some such way as two
magnets do when opposite poles are presented to each other; or
two charges of electricity of opposite nature. In fact, there is no
doubt that both magnetic and electric attractions are at work. We
are not entirely ignorant of their mode of action, but we know
much more about the rules of combination—that is to say, about
the facts of chemistry—than we do about the details of the attractions. However, we need not trouble ourselves about these matters
for the present; we have merely to realise that there are forces
drawing atoms together.

We may now ask why, if there are such forces, the atoms do not
all join together into one solid mass? Why are there any gases

or even liquids? How is it that there are any atoms at all which do not link up with their neighbours? What prevents the earth from falling into the sun and the final solidification of the entire universe?

The earth does not fall into the sun because it is in motion round the sun, or, to be more correct, because the two bodies are moving round one another. It is motion that keeps them apart; and when we look closely into the matter we find that motion plays a part of first importance in all that we see, because it sets itself against the binding forces that would join atoms together in one lump. In a gas, motion has the upper hand; the atoms are moving so fast that they have no time to enter into any sort of combination with each other: occasionally atom must meet atom and, so to speak, each hold out vain hands to the other, but the pace is too great and, in a moment, they are far away from each other again. Even in a liquid where there is more combination and atoms are in contact with each other all the time, the motion is so great that no junction is permanent.

In a solid the relative importance of the attractive forces and the motion undergoes another change: the former now holds sway, so that the atoms and the molecules are locked in their places. Even in the solid, however, the atoms are never perfectly still; at the least they vibrate and quiver about average positions, just as the parts of an iron bridge quiver when a train goes over it. It is difficult to realise that the atoms and molecules of substances which appear to be perfectly at rest, the table, a piece of paper, the water in a glass, are all in motion. Yet many of the older philosophers grasped the fact. For example, Hooke, an English physicist of the seventeenth century, explains by a clear analogy the difference which he sup-posed to exist between the solid and the liquid form: ascribing it to a movement of the atoms which was greater in the liquid than in the solid state. "First," he says, "what is the cause of fluidness? This I conceive to be nothing else but a very brisk and vehement agitation of the parts of a body (as I have elsewhere made probable); the parts of a body are thereby made so loose from one another that they easily move any way, and become fluid. That I may explain this a little by a gross similitude, let us suppose a dish of sand set upon some body that is very much agitated, and shaken with some quick and strong vibrating motion, as on a millstone turn'd round upon the under stone very violently whilst it is empty; or on a very stiff drum-head, which is vehemently or very nimbly beaten with the drumsticks. By this means the sand in the dish, which before lay like a dull and unactive body, becomes a perfect fluid; and ye can no sooner make a hole in it with your finger, but it is immediately filled up again, and the upper surface of it levelled. Nor can ye bury a light body, as a piece of cork under it, but it presently emerges or swims as 'twere on the top; nor can ye lay a heavier on the top of it, as a piece of lead, but it is immediately buried in sand, and (as 'twere) sinks to the bottom. Nor can ye make a hole in the side of the dish, but the sand shall run out of it to a level. Not an obvious property of a fluid body, as such, but this does imitate; and all this merely caused by the vehement agitation of the conteining [*sic*] vessel; for by this means, each sand becomes to have a vibrative or dancing motion, so as no other heavier body can rest on it, unless sustein'd by some other on either side: nor will it suffer any body to be beneath it, unless it be a heavier than itself."

Hooke's experiment can be repeated in a somewhat different form. A cylindrical metal box, ten inches wide and three inches deep, is fixed upon a platform which is supported on metal balls so that it moves easily. It is connected through an eccentric joint with a turning table as shown in Plate I B. When the wheel is turned rapidly, the box and the sand which it contains are violently agitated as Hooke prescribes. The details of the mechanism are best understood by reference to the figure. A heavy metal ball placed on top of the sand disappears at once, and light objects, such as ping-pong balls, rise to the surface. A very ludicrous effect is produced if we

PLATE I.

A. Model of bismuth crystal.

The model shows the arrangement of the atoms in the bismuth crystal. Each atom is represented by a round ball with three flats upon it.

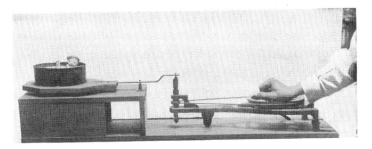

B. The shaking sand box.

The circular box is on the left; celluloid figures just showing.

bury in the sand some of the celluloid figures which cannot be made to lie down because they are heavily weighted at the bottom. The figures slowly rise out of the sand and finally stand erect. (Plate I B and Fig. I A.)

W. H. Bragg

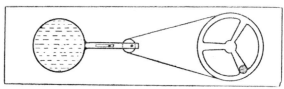

Fɪɢ. ɪ, A.

We know now that the motion of the atoms of a body is really its heat: that the faster they move or vibrate the hotter the body becomes. Whenever we warm our hands by the fire, we allow the energy radiated by the fire to quicken up the movements of the atoms of which the hands are composed. When we cool any substance we check those movements. If we could still them altogether we should lower the temperature to a point beyond which it would be impossible to go, the absolute zero, as it is usually called, 273 degrees centigrade below zero.

As I have said already, we have found two new allies, radioactivity and X-rays, in our attempt to see the very minute atom. They have increased the fineness of our vision some ten thousand times. The microscope had done its best for us; but the smallest thing which it could show us was composed of billions of atoms. No improvement could be made in the microscope lenses: technique had reached its highest. The difficulty was really due to the fact that light is a wave motion and light waves cannot show us the details of objects unless the objects are much larger in every way than the length of the wave. We wanted a new light of very short wave length. It came in the form of the X-rays. At the same time radioactivity came to show us what a single atom could do by itself if it were given a tremendous speed. We can now both see the single atom, indirectly no doubt but quite usefully, and also observe something which it does: the X-rays help us with the former and radioactivity with the latter. I hope to explain to you how both these agents are adding to our knowledge, and I will take radioactivity first.

The atom of radium might be roughly represented in size by one of the larger balls that lie before you. It is one of the heaviest and largest of the atoms; a number of them together form a substance which is a metal like iron or gold. It is, of itself, in no obvious way peculiar as long as it continues to be an atom of radium, but, for some reason, which no one understands, there comes a moment when it bursts. A small portion is hurled away like the shot from a gun, and the remainder recoils like the gun itself. The remainder is not radium any more, it is a smaller atom, having entirely different properties. The radium has turned into a new substance. As a matter of fact, the new substance is a gas, while the projectile turns out to be an atom whose weight is low down in the series of atomic weights, the lowest but one in fact; it is called helium. No one knows what brings about the explosion, nor does any one know a way of hastening it, or of hindering it. The radium atom is just as likely to explode at any given moment if it is in a furnace as if it is immersed in liquid air. Indeed, its independence of its surroundings in respect to its time of explosion is shown in a much stronger light by the fact that combination with other atoms makes no change. Combination, or molecule-forming, is, no doubt, concerned with the outside arrangements of the atoms, but the bursting of the atom comes from inside.

The old alchemist tried to find a means of converting one atom into another, preferably lead into gold. In the action of radium there is a transmutation, to use an old word, of the kind of which the alchemist dreamt. But it is not exactly what he strove for, in two ways. In the first place, it cannot be controlled by human will—which is extraordinary, because there are not so many things of which this can be said. Even when an operation is quite beyond our power to understand, we can often decide whether or no it shall happen. We cannot understand how a seed germinates, much less make one that will do so; but we can lock up seeds in a drawer and prevent them from germinating as long as we like. But the radium explosion does not wait on anything which we do.

In the second place, the transmutation does not end in gold: it ends rather in lead. The gas which consists of atoms of radium that have shot off one atom of helium is very short-lived: the average life of each of its atoms is a little less than four days, in contrast to the average life of the radium atom, which is about two thousand years. The second explosion "transmutes" the gas atom into a new substance called Radium A, and on the occasion another helium atom is shot away. There is a further succession of explosions, at very varying average intervals, and the final product is actually lead, not gold. The gas was called the "radium emanation" by Rutherford, who discovered it.

The whole operation is very wonderful, but I want to call attention to what happens to the projectile when it has left the gun. The velocity with which it starts is so great that one could never have thought any particle of matter could have possessed it. When Huyghens argued with Newton on the subject of the nature of light, he condemned Newton's idea that light consisted of a flight of corpuscles, on the ground that material particles could not possibly travel as fast as light had just been found to move. It is curious that we now find atoms moving with speeds comparable with, a tenth or twentieth of, that which then seemed impossible. There are even certain particles, called electrons, also emitted by radioactive substances, which travel, in some cases, very nearly as fast as light. It is also curious that the second argument of Huyghens was equally unfortunate in view of the observed phenomena of radioactivity. He said that it would be impossible, on Newton's theory, for two people to look into each other's eyes because the particles would meet each other and fall to the ground. We shall presently see that this argument also is set at nought by the facts of radioactivity.

The velocity with which the helium atom begins its flight is something like 10,000 miles in a second. In less than a minute it could get to the moon and back again if the speed were maintained, but the curious thing is that for all the speed and energy with which it starts it never gets far when it has to pass through anything material. Even if it is allowed to finish its course in the air, its speed has fallen to something of quite ordinary value after it has traversed a course of two or three inches in length. The course is, in general, perfectly straight, as we shall presently see in an actual experiment, and this is the very important point which we must consider with particular care. At first sight one does not realise how remarkable it is that its path should be *straight:* one thinks of a bullet fired through a block of wood, let us say, and making a cylindrical hole, or of the bullet in its straight course through the air. But the comparison is unfair. The bullet is a mass of lead enormously heavier than any molecule which it meets, and it brushes the air aside. But the helium atom is lighter and smaller than the atoms of nitrogen or oxygen of which the atmosphere is mainly composed, and we must think of some more truthful comparison. Suppose that a number of billiard balls are lying on a billiard table, and let them represent air molecules. If they are in movement the picture will be more correct, but the point does not really matter. Now let us drive a ball across the table aiming at a point on the opposite cushion, and watch what happens as the ball tries to get through the crowd that lies on the table, which crowd may or may not be in movement. It hits one of the balls and is turned to one side; it hits several in succession, and soon loses all trace of its original direction of movement. Shall we now drive it with all the force we can, and see whether it keeps any more nearly to the straight path? We try, and find that there is no improvement at all. The straight path cannot be obtained by any increase of speed, however great.

This picture or model is much more faithful than that of the moving bullet, and shows more clearly the remarkable nature of the radium effect. A helium atom must encounter a very large number of air molecules if it proceeds on a straight-line path, and if the atoms are of the size we have supposed them to be. In fact, the molecules lie far more thickly on the path than we can represent by the billiard-table model. It is possible to calculate how many air molecules, some oxygen, some nitrogen, would be pierced by a straight line three inches long drawn suddenly at any moment in the air, and the result is to be expressed in hundreds of thousands. How can the helium atom charge straight through this crowd, every member of which is heavier than itself? It does so, however, and we have to find some explanation.

Perhaps it might be thought that the straightness of the path is only apparent, and that if we could look into it in sufficient detail we should see that it was made up of innumerable zigzags made in going round the molecules met with. But a moment's reflection shows that the idea is absurd: the atom would need to possess the intelligence of a living being to give it the power of recovering a line once lost. If there were a cake shop on the opposite side of a crowded street, and if we gave a boy sixpence and directed him to the shop, he would no doubt pursue a path which was effectively straight, though it would be broken up by the need of dodging the various people and vehicles which the boy met with. But one cannot imagine an atom of helium doing anything of the sort.

There is only one way of explaining the marvel of the straight path: we must suppose that the helium atom *goes through* the molecules it meets, and that somehow it is enabled to do so by the fact that it is moving at such an unusual speed. It is a very startling idea. However, no other suggests itself; and, as a matter of fact, it

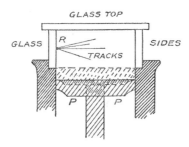

Fɪɢ. 2.—Section of the expansion chamber in Mr. C. T. R. Wilson's apparatus for measuring the track of helium atoms (see also Plate II).

The piston *PP* is dropped suddenly from the position indicated by the dotted lines to the position indicated by the full lines: so that the air in the chamber is suddenly chilled by expansion and fog settles on the tracks of the helium atoms shot out by the radium at *R*.

turns out that we can explain many other things by its aid. Consequently, we feel sure that we are on the right track.

It is time now that we should see this effect with our own eyes: the conclusion at which we have arrived is so new and so full of meaning that we would like to have an experimental demonstration if possible, and convince ourselves of the reality of these straight-line paths. We owe to Mr. C. T. R. Wilson a beautiful piece of apparatus which gives us a vivid picture of what happens, and we will make use of it at once. The experiment is, in my opinion, one of the most wonderful in the world of science. We are going to see the actual tracks of separate helium atoms, each of which begins its course at a speed of ten thousand miles a second and yet completes it after traversing about three inches of air. But we must first enter upon some explanation of how the apparatus works; for there are ingenious devices in it.

There is a cylindrical box of brass, with a glass top and a base which can be raised or lowered so as to alter the depth of the box. There is a machinery of wheels, cranks and levers by which the bottom of the box can be suddenly dropped at convenient intervals. Whenever this happens, the air or other gas which the box contains is chilled by the sudden expansion. We shall study effects of this kind more carefully in the next lecture. At the side of the box, in its interior, a minute speck of radium is mounted on a suitable holder. Every moment some of its atoms break up and expel atoms of helium, of which a certain number are shot straight into the box. The diameter of the box is big enough to allow the atoms to finish their courses in the air within. The average life of radium is so long that even if the apparatus held together for two thousand years, half of the radium speck would still be left. Yet each second, ten, twenty or a hundred atoms disappear in the expulsion of the helium atoms. Perhaps in no better way can it be shown how many atoms are concentrated in a small compass.

The air in the chamber is kept damp, consequently the chill due to expansion tends to produce a fog. Fog when it has to settle prefers to deposit itself on a solid nucleus of some sort, rather than to form independent drops in the air. The small particles of dust, if there are any, are made use of, which is the reason why fogs so readily form in a dirty atmosphere. But of all things moisture prefers to settle on those atoms through which the helium atom has passed. The reason is that the atom is temporarily damaged by the transit: a small portion has generally been chipped away. The portion removed is what we now call an "electron"; it is charged with negative electricity, and the atom which has lost it is correspondingly charged with positive electricity. The electron set free settles on some neighbouring atom, sooner or later; and in consequence there are two charged atoms, one positive and one negative, where previously there were no charged atoms at all. The charged atoms have a great attraction for moisture, and the fog forms on them in preference to anything else. If, therefore, a helium atom has just made its straight road through the gas,

PLATE II.

[By courtesy of the Cambridge Scientific Instrument Co.]

Shimizu-Wilson ray track apparatus.

The appparatus which shows the tracks of the helium atom shot out by radium. The chamber as selected in Fig. 2 is to be seen on the upper left of the figure. The disc to the left of it is a screen, in which is a hole. The light from a lantern—not shown—shines through the hole and lights up the fog tracks. A second screen revolves, and lets the radium rays, *i.e.* the helium atom, shine into the chamber just before the expansion is made. On the right is the driving machinery.

and has left behind it numbers of charged atoms on its track, and if, at that moment, the sudden expansion causes a chill, fog settles along the track. A bright light is made to illuminate the chamber, so that the fog tracks are visible as bright straight lines, showing against the blackened background of the bottom of the cylindrical chamber. They last a few seconds, and then the fog particles slowly disperse. If the helium atom completes its track just before the fog is formed, the line is sharp and clear; because the charged atoms have not had time to wander from the track. But if the track is made some time before the expansion, the line of fog is more diffuse. It is to be remembered that the helium atoms are being shot out all the time, day and night; but it is only when an expansion is made that tracks are made visible.[3]

If we watch the successive expansions, we see that the tracks, though quite straight over large parts of their course, do undergo at times sudden sharp deflection, especially when they are nearing the end. This remarkable effect turns out to be most important, and we must refer to it presently.

Let us now try to picture to ourselves in what way we must modify our first conception of the atom so that we can explain the effects we now see. The atoms must be so constituted that when they meet one another in the ordinary way, as, for example, when molecules of oxygen collide in the atmosphere, they behave as if each had a domain of its own into which no other might enter. Or, when they are pressed together, as in a solid, they occupy as a whole an amount of space which is sufficient to make room for them all. But when one atom—the helium atom is our chief

---

[3] In the lecture the working of the apparatus was illustrated by a kinematograph film which had been made for the purpose. It showed a series of successive expansions, each forming a new set of lines like those shown in Plate III.

PLATE III.

B

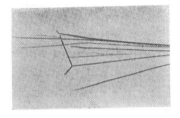

D

Alpha ray tracks.

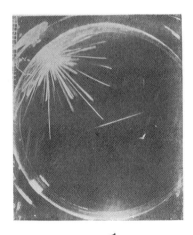

A

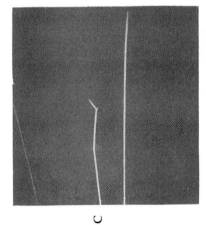

C

example—is hurled against others with sufficient speed, the one atom goes through the other, as if the defences round the domains had been broken down. We find a satisfactory explanation when we imagine each atom to be like a solar system in miniature. There is to be a nucleus, corresponding to the sun, and round the nucleus there are to be satellites or planets, which we call electrons. The nucleus is charged with positive electricity; each electron is charged with negative electricity, and all electrons are alike. The positive charge on the nucleus is just enough to balance the united negative charges of the electrons. The electrons are supposed to be in movement, just as the planets are revolving round the sun, but the movements are no doubt complicated, and their nature need not for the moment concern us at all.

Instead, therefore, of a round hard ball of a certain size, which was our first rough picture of an atom, we have something like a solar system in miniature. We can at once see how one atom of this kind can pass through another, just as we might imagine one solar system passing through another, without injury to either provided that no one body of one system made a direct hit on a body of the other and that the motion was quick enough. The latter condition is necessary because if one solar system stayed too long inside or in the neighbourhood of another there would certainly be very serious disturbances of the courses of the planets.

But then, we may ask, how can an atom, if this be its nature, have the power of keeping another outside its own domain? How can it appropriate any portion of space to itself, and prevent the intrusion of another atom when the speed at which they meet is low? The explanation becomes clear when we consider the special arrangement of the positive and negative charges. Every atom is surrounded by a shell or cloak of electrons; and, when two atoms collide, it is their shells which first come close together. Since like charges of electricity repel one another, the two atoms will experience a force which tends to keep them apart: in other words, they will resist encroachment on their own domains. This is, no doubt, a very rough picture of what actually happens, and as a matter of fact it is difficult to explain the strength of the resisting forces on such a simple hypothesis. Still, it is on the right lines, no doubt. When the two atoms approach each other at a high speed, the system of electrons and nucleus of one atom slip through those of the other. A model will help to illustrate the point.

Plate IV A shows a set of bar magnets mounted on spiral springs and standing erect. The top of the inside magnet is a north pole, and the tops of the magnets of the outside ring are south poles. The model represents roughly the central nucleus surrounded by a ring of electrons. In the model everything is in one plane; in the atom it is not so, but the point is not important. A single magnet is suspended by a long thread from a point vertically over the "nucleus" magnet. Its lower end is a south pole and the length of the thread is such that the swinging magnet just clears the fixed magnets.

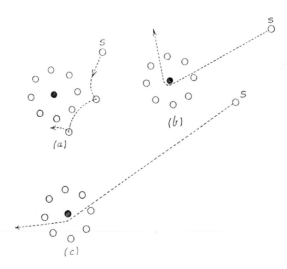

FIG. 3.—(*a*), (*b*) and (*c*).

PLATE IV.

A. Bar magnets on spiral springs.

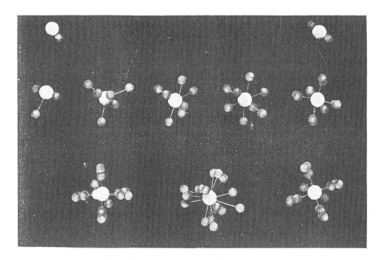

B. Models of atoms with electrons.

Observe now that if we pull the swinging magnet to one side (*S* in Fig. 3, *a*), but not too far, it moves towards the fixed set and is unable to enter in. It seems to knock at the door at one place after another, but always recoils. Just so would an electron beat in vain against the outer defences of an atom, if it did not beat hard enough. We can easily imagine that if the single swinging magnet were replaced by a system of magnets, like our stationary set, the same result would follow. Here we have a picture representing our atoms, as we now think of them, beating against each other and recoiling; each occupies a certain domain of space and prevents the intrusion of any other atom.

But if the swinging magnet is drawn sufficiently far to one side so that it acquires a greater speed than before, by the time it reaches the stationary set its momentum will carry it through. If the speed is very great, it shows no appreciable change in its motion due to its passing (Fig. 3, *c*); if the speed is rather less, it often suffers in going through (Fig. 3, *b*). It comes out less vigorous than when it went in; often it has changed the direction of motion also, and it has obviously left energy behind, for the magnets of the stationary set are left quivering. This happens no matter which pole of the swinging magnet is the lower, and clearly the same effect would be shown if the single swinging magnet were replaced by a more complicated set of nucleus and attendant satellites.

The behaviour of the model helps us to anticipate what we should find when atoms of our new design come across one another. If they approach at a moderate speed, they may rebound from one another; at a high speed they go through each other, and the higher the speed, the greater the chance of a passage without any obvious result. But there is always the chance that the nucleus of the moving atom may go so near to the nucleus of the atom through which it is passing that it experiences a perceptible deflection. The smaller the nuclei are, the less likely it is that this will happen.

You will have guessed already that you have actually seen such deflections as these in the kinematograph picture, such as also are shown in Plate III A to C. The tracks of the helium atom are quite straight in the main, but there are decided breaks in the straight lines, usually not more than one or two in each track. They are found mainly towards the finish. This is what might be expected, since the motion will then be slower. Several of them appear in Plate III A; a very good example of this kind of track is reproduced on a large scale in Plate III C. The upper track shows a slight but sharp deflection at a little distance from the end of its course, and a more pronounced deflection further on. Nearly every track shows some deviations at the very end. Thus the new conception of atomic structure explains all the effects in a satisfactory way.

It is strange to think of an atom as being empty as a solar system: not a round, hard and absolutely impenetrable body, but a combination of nucleus and electrons which occupies a certain space somewhat as an army occupies a country. The bodies of the soldiers do not fill the country from boundary to boundary; but enemy soldiers may not enter nevertheless.

These very characteristic pictures are the fruit of much watching and photographing. Breaks are found at every expansion, but it may be necessary to wait for a really good one. A very fine picture is shown in Plate III D. It is due to Mr. P. Blackett. In this case, helium was used instead of air. The nucleus of the flying helium atom, in traversing a helium atom belonging to the gas, has made an almost direct hit on the nucleus of the stationary atom: it has cannoned off it, as a billiard player would say. Both atoms now move with not unequal speeds, and both make fog tracks, as the figure shows. In Plate III C if we look carefully, we see that there is a minute spur on the last bend of the track already mentioned, which means that in this case an atom of oxygen or nitrogen has deflected the helium atom and has recoiled in consequence. Its track is very short, because it is much heavier than the atom which struck it, and, therefore, the velocity given to it has been comparatively small.

There is a certain curious feature to be found in some of the photographs which may well be explained. In some of the tracks there are gaps, as if the fog settling had failed. This is indeed the actual fact: there is no moisture to settle, because a helium atom has gone that way some very short time before and has used up the moisture in the neighbourhood. In Plate III B several tracks due to radium emanation are shown. They seem to start from anywhere in the chamber because the atoms of the emanation have wandered about the chamber before blowing up.

The next question that arises is as to the number of electron satellites which each atom possesses. Here we come to a very beautiful and remarkable feature of the new discoveries. It is not necessary to explain in full how it was discovered; we will be content with describing it.

In the atom as we now have it the nucleus is charged with positive electricity, the amount of the charge being just enough to neutralise the negative charges on the attendant electrons. All electrons, as we have already seen, are alike. We find that atoms differ in the number of attendants which they can maintain, and that the statement of that number describes the atom completely so far as its attitude towards other atoms is concerned. For instance, the atom of carbon can hold six electrons; the positive charge on the nucleus is the counterpart of six standard negative charges. Every atom which can retain six electrons is a carbon atom: no other definition of the carbon atom is required. Just so the "seven-electron" atom is nitrogen, the "eight-electron" is oxygen and so on. All numbers are found in nature, with very few exceptions, from the "one-electron" atom—hydrogen—up to the "ninety-two-electron" atom—uranium. The missing numbers will probably be found some day; more or less accidentally, it may well be.

PLATE V.

A. Floating magnets.

When the number of floating magnets is small, they form into a single ring, but when the number is increased, they form concentric rings.

B. Crystals in tube containing emanation.

(From Prof. F. Soddy's " Interpretation of the Radium " (John Murray), by the kind permission of author and publisher.)

We may use models as a rough illustration of the point. The nucleus (Plate IV B) is represented by a white ball of solid rubber, the electrons by smaller balls forming the heads of pins which are stuck into the centre ball. The pins may be of different lengths (p. 77) [not included here].

It is strange that the immense variety in Nature can be resolved into a series of numbers. It was at one time thought that the various sorts of atoms owed this variety to something more than that; it is a great surprise to find such a simple kind of difference between atom and atom. The unchanging feature of any particular sort of atom is the positive charge of electricity on the nucleus. It is in consequence of this that the proper number of electrons gather round. We may expect that they will arrange themselves in some fashion; we shall see later that they certainly do so. The sort of arrangement they take in each case, and the nature of the forces put into it, are very difficult questions, most of which we may well put to one side for the present, contenting ourselves with one or two simple aspects of the problem.

In the first place, it is interesting to watch the assembling of the little vertical magnets floating in the glass tank (Plate V A). They are buoyed up by ping-pong balls, painted black, and, so that we may see them easily, they carry white ping-pong balls at top. The magnets are all the same way up, so that naturally they repel one another and cluster round the edge of the basin. But there is an electromagnet underneath the bowl; which, when made active, draws the small magnets together. The arrangement in which they settle finally is governed partly by the pull towards the centre and partly by the mutual repulsions. Something of this kind must take place in the atom, but we must not push the analogy too closely, because the forces may be quite unlike those which are exerted in the model. We must content ourselves with observing that when there are only a few magnets afloat they group themselves in a ring; but when the number is increased they arrange themselves in concentric rings. A pretty effect is produced by putting in each additional magnet at the edge of the basin and watching it float away in a stately fashion to take its proper place.

A similar division into concentric shells or groups is found in the arrangement of the electrons round the central nucleus of the atom.

We will consider this more carefully in the next lecture. The experiment does not prove that there ought to be such an arrangement, but certainly suggests it.

We may now see more clearly what happens when the helium atom injures the atoms through which it passes and renders them attractive to the particles of moisture that form the fog. It is possible, in fact, for an atom to be deprived of one of its attendant electrons. Having lost one, it resists more strongly the loss of a second, still more of a third. As the helium atom goes on its way, it strips one atom after another of an attendant, and the electron set free goes off on a course of its own. But its separate life is very short-lived: it is soon attached to another atom. The atom that has lost an electron is now positively charged; the gainer is negatively charged. The two atoms would make things even again if they came sufficiently close together, and as they move about in the gas the negatives and the positives do in the end give and take electrons, and the whole gas is neutral once more.

There is a beautiful experiment with which we may end this lecture. When the helium atoms strike certain substances they excite a phosphorescent glow. It is really, when we look into it closely, a set of minute flashes due to the impacts of the separate atoms; under a microscope the effect is as when we drop pebbles into a phosphorescent sea. The glass vessel (Plate V B) contains crystals that phosphoresce under the stimulus of the swift-moving helium atoms; one is kunzite, another zinc sulphide, another willemite. In another tube is a quantity of radium emanation: the gas which, you will remember, is the immediate descendant of radium itself. When it is released and is allowed to pass into the tubes containing the crystals the latter glow in brilliant colours. In the figure the crystals have been made to photograph themselves by their own phosphorescence.

The radium action has, we see, given us a remarkable insight into the structure of the atom, for which there is a general reason to be given. The student of science has long been familiar with the existence of various atoms and with their properties; he has never seen one, nor the effects of one. He has handled atoms in crowds only. When the chemist causes elements to form compounds, or analyses compounds into elements, he deals with enormous numbers of atoms in any operation big enough to see. But in this radioactive effect we observe the action of one atom at a time, and here lies the secret of the advance. The speed of the helium projectile, a hundred thousand times the speed with which the atoms move ordinarily when they form part of a gas, gives the individual atom the power of making itself felt. When we look at the fog tracks, we see the actions of separate atoms; we see something which would have filled the early defenders of the atomic theory with astonishment and pleasure. One atom of helium passes through one atom of oxygen, let us say; and comes out on the other side, and both may bear evidences of the encounter. Effectively we use such evidence to help us to determine the nature of the atoms. The helium atom is like a spy that has gone into a foreign country and has come out again with a tale to tell.

Our Electrical Supply

William Lawrence Bragg*

IN the first three chapters we have been studying the principles of electricity and magnetism, becoming familiar with the behaviour of charges, currents, and magnetic fields. The remaining chapters deal with the practical application of these principles to three great divisions of electrical engineering. In the first place there is the generation transmission and use of electrical power, which comes under the heading of heavy-current electrical engineering. Then there is the use of the electrical current to convey messages from one place to another, by telephone, telegraph or submarine cable. The currents employed are very small, and our concern is not with the development of power but with delicate and complicated variations in the electrical current. Just as paint may be used either to cover the Forth Bridge or a canvas in the Royal Academy, so the electrical current may be used to convey either gross power or a message from one human brain to another. This latter is light-current electrical engineering. The final chapter describes some properties of high-frequency currents. It would form a natural introduction to the study of wireless telegraphy and telephony, but that subject is so vast that it could not be dealt with comprehensively here and

---

* The original version appeared in: William Lawrence Bragg, Chapter 4, *Electricity*, London, 1936, pp. 143–192.

I must be content with a description of the principles of the oscillating electrical circuits on which wireless is based.

The present chapter deals with the generation and transmission
of electrical power on a large scale.

1. THE FUNCTION OF A SYSTEM OF ELECTRICAL SUPPLY

A system of electrical supply is not a new source of power, it is
merely a new way of conveying power from one place to another.

Previous to the industrial development which began about 150
years ago, most of the power which was required for various
human activities was developed in the muscles of men or of animals. Men worked the land themselves, or used horses or oxen to
draw the plough. Transport was by horse-drawn carriages or pack
horses in the more advanced communities and by human porters
in more primitive ones. In earlier times much heavy work was
done by gangs of slaves, who could be fed on the cheapest of food
and whose bodies turned that food into available power. The typical
exceptions to the almost universal use of muscle were such simple
operations as could be carried out monotonously and in a leisurely
way at a fixed point such as grinding corn or pumping water.
Windmills and water-wheels were used for grinding corn because
they could be placed at convenient sites on hills or in valleys, and
the corn could be brought to them. Windmills were used for
pumping water in flat countries like Holland and our fen district in
Norfolk. They were only required to keep down the general water
level in the ditches and had no sudden emergencies to cope with.
A spell of calm weather when the windmills could not work was
compensated for by a succeeding windy spell.

The development of the steam engine marked an immense
change, for it became possible to have a steady supply of power at
any place to which fuel could be brought. As one can observe in
Lancashire, the earliest factories had to be placed in valleys where

water power was available. With the advent of steam power, they could be placed anywhere so long as coal could be brought cheaply to that place, and it was for this reason that the big industrial areas in England grew up around the coal-mines. Even in these days of railways, it is still an expensive business to convey coal for long distances, as is shown by the difference in price of coal in different parts of the country.

The development of the generation and transmission of energy in electrical form marks another immense step forward in control over power. It must be emphasized again, however, that this is not because a new source of power is available, but because a new way has been found of sending power from the places where it is conveniently developed to the places where it is needed. I may perhaps take an analogy from the way power is distributed in a large workshop or mill. A central engine drives a system of overhead shafts with pulleys, from which belts descend to many machines. It would be highly wasteful and inconvenient to have a small engine to drive each machine, and this difficulty is overcome by transmitting power through the shaft and belt drive. We must compare a system of electrical supply to the system of shafts and pulleys; for it is essentially a method of transmission. It is still necessary to have a source of power or 'prime mover' such as a steam engine. This engine drives a dynamo in a power station which converts mechanical power into electrical power. The current flows out from the station along the mains and is used to drive motors at the places where mechanical power is needed, or to provide us with light and heat. It is a marvellous method of transmission because it is so simple and efficient. Instead of a clumsy system of belts and pulleys or of cog-wheels which can only transmit the power a few yards and even then waste a good deal of it, we can convey many thousand horsepower along slender wires. I believe it was Oliver Wendell Holmes who called it power stripped stark naked. The power can be sent hundreds of miles by transmission lines, and then run by cables to any part of a building where it is required. There is a story of a Lancashire manufacturer which illustrates its possibilities very well. An electrical engineer was trying to persuade him to convert his factory from the belt drive with which he had grown up into an electrical system, and explained that if the machines were driven by motors they could be arranged in a more convenient way than if they all had to be aligned so as to be driven by overhead shafts. At last the manufacturer grasped the point and said to the engineer: 'Do you mean to tell me that this electricity can go round a corner?'

You can understand, then, what a revolution the widespread distribution of electrical power is bringing about. Instead of having to build an engine at any place where power is required, we only have to link that place to the main electrical supply by overhead wires or underground cables, the power being developed wherever it is most convenient. Factories need no longer be grouped around the coal areas nor need each factory have its chimney belching forth smoke.

A big generating station can afford to instal the plant necessary to cope with the smoke from its furnaces, cleaning it and removing the harmful gases. By centralizing the development of power in power stations, it is also possible to develop it more cheaply, because much more energy is got from each pound of coal by large engines than by small ones. Electricity is one of the greatest assets in curing the ugliness, dirt and lack of planning which have spoilt so much of the country and are bad legacies from the last century.

2. ELECTRICAL UNITS

I have tried to avoid definitions and formulae in this book as in the lectures, but since we are now discussing practical electrical engineering we must have some idea of the units used to measure electrical quantities, just as we have to know roughly what is meant by 'second,' 'hour,' 'foot,' 'mile,' 'pound,' and 'ton,' in the ordinary practical affairs of life when we are ordering the coal or reckoning how long a motor journey will take.

A good way of acquiring a familiarity with electrical units is to quote figures for some examples with which most people are familiar.

The *ampere* is the unit of *rate* of flow of current.

The current flowing through an ordinary 40-watt lamp on the 230-volt mains is about $\frac{1}{6}$ ampere. An electric radiator uses from 10–20 amperes and it is therefore necessary to use much stouter leads for wiring up a radiator than those which are required for a lamp. The self-starter of a motor-car requires as much as 100 amperes for the few seconds it is in action, and frequent attempts to start a cold engine soon run the battery down. The minute currents which convey our voices along telephone wires are of the order of one ten thousandth of an ampere.

The *Volt* is the unit of potential, or as it is sometimes expressed, of 'electromotive force' (E.M.F.). You will remember that if we compare an electric current to the flow of water, we may say we measure the rate of flow in amperes and the pressure or 'head' in volts.

A dry cell has an E.M.F. of 1.5 volts, and an accumulator cell of just over 2 volts. A motor-car battery generally consists of six accumulator cells in series, giving a total of 12 volts. The standard voltage for house supply in this country is 230 volts, though as the current supply is of the 'alternating' type, to be described below, this figure requires further definition. Alternating current is generated in power stations at various voltages, but 6,600 volts, 11,000 volts and 33,000 volts are common. The 'Grid' (see page 181) [p. 273 here] operates at 132,000 volts. Potential differences in a thunderstorm rise to one thousand million volts.

The *Watt* is the unit for measuring power, or the rate at which electrical energy is being used or generated.

To obtain the number of watts, we multiply volts by amperes.[1] The volts measure the pressure which is driving the current, and the amperes the rate at which current is flowing, so clearly they must be multiplied if we are to get a measure representing the power. Lamps are marked with the number of watts they consume, 25, 40, or 60 watt lamps being common ratings for household use. When a measure of larger amounts of power is required, the kilowatt is convenient. A kilowatt is one thousand watts, and corresponds to $1\frac{1}{3}$ horse-power.

The *Coulomb* is a measure of electric charge, being the amount which would pass any point in a circuit each second if an ampere were flowing. We have previously mentioned 'electrostatic units of charge'; the coulomb is a more convenient unit when dealing with currents. An accumulator cell delivers during discharge a total number of coulombs which depends on the size of its plates. Since this number is inconveniently large, we generally reckon the capacity of the cell in 'ampere-hours' and not coulombs which are 'ampere-seconds.' A small cell in a wireless set may have a capacity of 25 ampere-hours, and a cell in a motor-car of 100 ampere-hours.

The 'unit' which appears in the quarterly electricity bills represents the supply of one kilowatt for an hour or its equivalent. An electric radiator consumes two or three kilowatts, say four horsepower. It brings home very clearly the cheapness of electrical power, if we think that when an electric fire is turned on we have the equivalent of several horses working for us, and that it only costs a few pence to hire them for an hour. A big power station like Battersea will yield 480,000 kilowatts when fully developed, and work continuously as hard as more than half a million horses. If we allow a reasonable working day to the horses, we might reckon

---

[1] It is again necessary to qualify this statement in the case of alternating current, where a relation called the 'power factor' has to be considered.

that it would take two million to provide us with the equivalent amount of power.

The *Ohm* is the unit of resistance. Everyone who has ventured at all into the study of electricity is familiar with Ohm's Law, which states that the current passing through a conductor is proportional to the electromotive force driving it. The ohm is so chosen that $C = E/R$ where C is measured in amperes, E in volts, and R in ohms. Ohm's Law is almost unique in Physics, because it really is true! In the case of most 'laws' we have to make many qualifications and corrections, but within the accuracy of measurement Ohm's Law may be relied upon.

The intensity of the shock when a discharge passes through our bodies depends on the magnitude of the current. The shock when the 230-volt mains are touched gives one a sharp prick, but it is often harmless because the skin when dry has a high resistance and this voltage is unable to drive a dangerous current through it. It is a very different matter when the skin is wet so that a really good connection is made. A shock from the 230-volt mains may then be very nasty or even fatal. One must be particularly careful not to have faulty switches, or exposed leads to apparatus such as an ultra-violet equipment, in a bathroom, where they may be touched by damp hands. A higher voltage than about 500 begins to be dangerous in any event.

3. ALTERNATING CURRENT

When one is buying electrical equipment, one is generally asked whether the local supply is 'direct current' (D.C.) or 'alternating current' (A.C.). Most generating stations in this country are now being connected up with the grid, and the supply for ordinary domestic use is being converted into a standard one of '230 volt 50 cycle A.C.' We must see what this means.

A direct current flows steadily in one direction, like the water in a stream; an alternating current flows backwards and forwards like the tides in an estuary. The mains which pass into our houses from a direct current supply (D.C.) have got a constant difference of potential between them. The potential of one may be, for instance, always 200 volts higher than that of the other, so that if we connect a lamp across the mains there is a steady pressure of 200 volts driving current through the lamp. The mains from an alternating current supply (A.C.) on the other hand have a fluctuating potential difference between them. First the one, then the other, is at the higher potential, and a current drawn from them flows alternately in either direction. If the current goes forwards fifty times and backwards fifty times in a second, we say that the supply has a *frequency of 50 cycles,* a cycle being a general name for one of a series of repeating events.

The nature of an alternating current supply can be illustrated by an artistic experiment, which we showed at the Christmas lectures, using a well-known effect to indicate the direction of the current. When a current runs through a solution of potassium iodide, it liberates iodine at the anode (see Fig. 33) [not included here] by electrolysis, and ordinary starch paste turns a violet-black colour when it reacts with small traces of free iodine. A large linen cloth is soaked in starch paste in which potassium iodide has been dissolved, and it is then spread while still damp on a sheet of tinned iron. (It is worth remembering if this experiment is being repeated that a zinc sheet is not suitable, because the zinc stains the cloth dark brown in a very short time by chemical action.) If now we take two metal rods, provided with insulating handles and connected to a D.C. supply of a few volts, and let the negative rod touch the metal sheet, a beautiful violet line can be drawn on the sheet with the positive rod because the iodine liberated by the current reacts with the starch. Fig. 69 (Plate 15) shows the author drawing a fish in this way. I strongly recommend this method of drawing to lightning artists, because an intensely dark and very uniform line follows the moving terminal, and can be produced at any speed with the lightest possible touch.

PLATE 15

Fig. 69. Drawing with a positive electrode on a sheet moistened with starch paste and potassium iodide. (*'Sphere' photograph*)

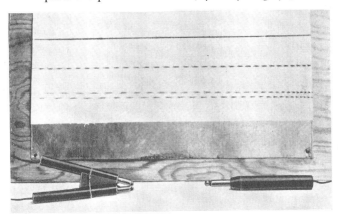

Fig. 70. The upper line is drawn with a positive electrode (direct current). The next line is drawn with one lead from an alternating current supply. The lower pair of lines are drawn with the two leads from an A.C. supply. Note the alternate spacing of the marks. The track was drawn in about half a second

If we exchange the terminals, only a faint smudge appears. The plate then becomes the anode and the terminal we are drawing with becomes the cathode. The iodine is liberated behind the sheet where it touches the plate, not in front of it, and the stain is only faintly seen through the sheet.

Fig. 70 (Plate 15) shows the effect of drawing rods attached to an alternating current supply across the sheet. In doing this experiment, the 230 volt A.C. supply may be used, provided one puts an electric lamp in each lead and care is taken to hold the rods by their insulating handles. If the lamps are there, an accidental contact between the rods only lights them, whereas if the leads are connected directly to the mains, the contact would make a 'dead short' and blow the fuses. The two rods are held side by side by their insulating handles, and drawn rapidly across the sheet. As you will see, each rod draws a dotted line, showing that it is alternately anode and cathode. Since when one is cathode the other is anode and vice versa, the dots are spaced alternately.

We now set a metronome to beat seconds, and after a little practice draw a pair of lines which start and stop at successive ticks of the metronome. The lines should be two or three feet long. We will find that there are about fifty dots on each line; there would be just fifty if it were possible to draw for exactly one second. This demonstrates the 50 cycle frequency of the alternating current.

One can get fine artistic effects by drawing with one rod (the other lead being connected to the metal sheet), and asking an assistant to switch on direct current or alternating current as the picture needs continuous or dotted lines. The only drawback is that these works of art are not permanent, for they fade in an hour or so.

An alternating current successively has a maximum in one direction, falls to zero, has a maximum in the opposite direction, and falls to zero again. Its average value is clearly less than its maximum value. By 'an alternating current of one ampere' we mean a current which has a maximum value of 1.41 amperes ($\sqrt{2}$ amperes) each way, one ampere being what is called its 'root mean square' value. The same holds for an alternating voltage. In a '230 volt A.C.' supply, the potential difference actually oscillates between ±325 volts.

An alternating supply of the kind we have been considering, which comes to us along two mains, is called a 'single-phase' supply. One of the cables coming into our houses is called the 'neutral' and is at zero potential. The other, the 'live' cable, has a potential oscillating between +325 volts and −325 volts as explained above. Although two wires run to each lamp, it is only necessary to have one lamp switch, connected to the 'live' main, because when the switch is off both wires become neutral and can be touched without getting a shock. On the other hand, if you look at the lines which take the grid current across country (see Fig. 96, Plate 27) you will notice *three* lines on insulators. (The fourth slender line linking the tops of the towers is not carrying current, it is merely an earthing line which connects them electrically.) These conductors are conveying 'three-phase' or 'poly-phase' current. If we call them 1, 2, 3, the lines reach their maximum positive potential in the order 1—2—3—1—2—3 fifty times a second. When 1 is at its maximum positive potential, 2 and 3 are negative and so on. Fig. 71 shows the potential changes of single-phase and three-phase supplies. Three-phase current is used for power transmission because it has certain technical advantages which will be described later, but the domestic supply is always single-phase.

4. WHY ALTERNATING CURRENT IS USED; TRANSFORMERS

What is the point of using alternating current? At first sight it seems a very odd thing to do. The cables going out from power stations are conveying alternating current. No sooner has the current started running in one direction than back it comes again. It is just as if the driving belts in a factory were going backwards and forwards instead of running steadily in one direction.

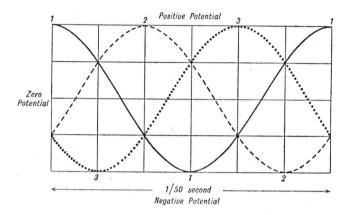

Fig. 71*a*. A three-phase supply. The potentials of the three wires are represented by the three curves.

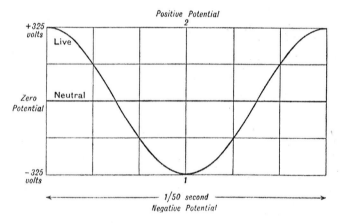

Fig. 71*b*. A single-phase supply. The curve represents the potential of one wire, the other 'neutral' wire being at zero potential.

To understand this, we must first consider the problem of transmitting electrical power from the generating station to the place where it is to be used. The cables which convey the current have a certain resistance, which is the greater the longer the distance the current has to be taken. Some of the power is wasted in driving the current through the cables against the resistance. The resistance is made as low as possible by using wires of copper or aluminium which have a high conductivity, and by making the cables thick. However, the weight of the cables cannot be increased beyond a certain point, for they then become very costly and too heavy to be carried as overhead lines by towers or poles. We must therefore avoid wasting power in the lines by keeping the current along them as small as possible, since the rate at which energy is lost in useless warming of the wires depends only on the resistance and the current. On the other hand, since vast amounts of power must be conveyed from one place to another, if the current is to be small the voltage must be very high, for you will remember that power is measured by multiplying current by voltage.

This is the sole reason for using high voltages on transmission lines. You cannot fail to have noticed the long strings of insulators on the towers, and the notices at the foot of each tower, 'Danger, 132,000 volts.'

Let us imagine we had to send the same electrical energy from one part of the country to another at 230 volts direct current, instead of 132,000 volts alternating current, with no more waste due to resistance than at present. Currents would be nearly 600 times as great, and the cables would have to be 35 feet across instead of $\frac{3}{4}$ inch as they actually are. You can see what an enormous advantage is gained by transmitting at the high voltage.

The energy is therefore sent out from generating stations at high voltages, in order to keep the transmission losses small. On the other hand, it must be tamed down to a small voltage before it can be used. The 230 volts of the domestic supply is practically the highest possible voltage which is safe in a house.

Alternating current is used because it is so easy to change from one voltage to another. In order to turn a *direct current* at one voltage into a direct current at another voltage, the first current must be used to drive a motor, which drives a dynamo so designed as to give the different voltage. This is a relatively expensive equipment and needs constant attention. It is also difficult to design motors and dynamos for very high voltages, because the insulation of the coils becomes a serious problem. An *alternating current,* on the other hand, can be converted from one voltage to another in a very simple way by a *Transformer.*

You may remember the story of the man who was shown a giraffe for the first time, and said, 'I don't believe it's true!' Of all the electrical apparatus, I think the transformer makes one most inclined to say the same thing when one first appreciates what it does. It seems almost incredible that anything so simple should be able to hand on energy in such a convenient way.

A transformer consists of two coils of wire wrapped around the same piece of iron. One coil is called the primary and the other the secondary. If an alternating current is fed into the primary, an alternating E.M.F. is set up in the secondary in which a current will run if the circuit is complete. Fig. 72 (Plate 16) shows the transformer in a very simple form. The primary is a coil wound round an iron bar, which projects for some distance beyond the coil. The secondary is a separate coil of wire, with its ends joined up through a lamp. If now we pass A.C. through the primary, and lower the separate coil over the iron rod, we shall find that the lamp lights up, faintly at first and more brilliantly as the secondary approaches the primary coil. You will notice that we are handing on a current from the one coil to the other without contact of any kind between the two.

The principle by which the transformer works is simply the principle of electromagnetic induction which we have already studied. The alternating current in the primary makes the iron rod an electromagnet with its North pole alternately at either end. Each time the magnetization of the iron rod is reversed, an electromotive force is induced in the secondary. We therefore get an alternating E.M.F. in the secondary which can be used to drive a current with the same frequency as that of the primary current.

If this were all, nothing would seem to have been gained, since we have merely got a current like that with which we started.

PLATE 16

Fig. 75. A transformer mounted on a pole.
(*Central Electricity Board*)

Fig. 72. The action of the transformer. The lamp is lit by currents induced in the coil connected to it, when an alternating current passes in the lower coil. (*'Sphere' photograph*)

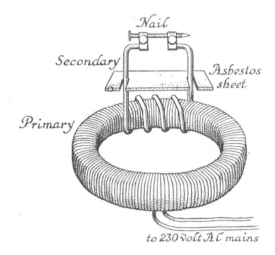

Fig. 73.   Melting a nail with a 'step-down' transformer. The asbestos sheet protects the transformer from the molten metal.

However, let us suppose that we make a transformer with a few turns of thick wire in the primary, and a large number of thin turns in the secondary. By passing a *large current at low voltage* into the primary, we are able to draw a *small current at high voltage* from the secondary. This will work backwards, for it does not matter which coil is called the primary and which the secondary. By passing a small current at high voltage into the coil with many turns, a large current at low voltage can be drawn from the coil with few turns.

This latter effect is shown by the experiment of Fig. 73. An iron ring is wound with a large number of turns and connected to the A.C. mains. A very thick copper wire is coiled two or three times around the ring as shown in the figure, and its ends are clamped to an iron nail. Putting quite a moderate current into the primary, so large a current flows in the secondary that the nail soon gets white hot and melts.

Fig. 74 (Plate 17) shows the opposite effect in a gigantic 'high tension transformer' used in the research laboratories of Metropolitan-Vickers to test insulators. The potential between one end of the

PLATE 17

Fig. 74.   A million-volt transformer. (*Metropolitan-Vickers*)

secondary winding and earth rises to one million volts, and produces the discharge seen in the foreground.

It is easy to see that as we increase the number of turns of wire in the secondary of a transformer, we increase the voltage. The magnetic changes produced by the primary induce a certain voltage in each turn of the secondary, and these voltages all add together. The rule for the increase or decrease is a simple one. If there are N times as many turns on the secondary as on the primary, the voltage goes up N times.[2]

A transformer thus gets its name because it is able to transform a supply of power represented by alternating current at a certain voltage into a supply at a higher or lower voltage. If used to raise the voltage it is called a 'step-up' transformer, if to lower the voltage it is called a 'step-down' transformer. It is so useful because it is relatively cheap to construct and has no moving parts. It does not need to be started up, and requires no attention when running. You may have noticed the iron huts, often to be seen near small towns or villages, with high voltage wires running to them. They contain transformers, converting the current from a high voltage to the low voltage which is supplied to the neighbouring houses, and they are only visited at intervals for a routine inspection.

Fig. 75 (Plate 16) shows a little step-down transformer mounted in a pole. The high tension supply at 11,000 volts is coming from the left and is led down to the transformer. One can see the low tension cables (230 volts) running away in the distance to supply local needs.

Fig. 76 (Plate 18) shows a large transformer used for converting three-phase alternating current[3] from a power station from 33,000

---

[2] We may emphasize here a point which is dealt with more fully below. Although we may get an increased voltage from the secondary, we cannot use it to produce an unlimited current, since we cannot get more energy from the secondary than is supplied to the primary. In other words what we gain in voltage we lose in current.

[3] It is actually three transformers in one, since the current is three-phase.

PLATE 18

Fig. 76.   A 30,000 kilowatt transformer for three-phase current, removed from its tank (*Metropolitan-Vickers*)

to 132,000 volts. This transformer can convert 30,000 kilowatts. The photograph was taken before the transformer was put in its case; when actually working it is placed in a large tank holding about 7,000 gallons of oil which serves a double purpose. It is a better insulator than air, preventing discharges between the coils at the high voltages for which the transformer is designed. The coil also cools the windings and the iron core, just as the cylinders of a motor-car are cooled by a water jacket. The transformer has big 'radiators' on either side also like car radiators, where the oil in turn is cooled by the air. You will see in the figure gaps in the windings of the transformer so that the cooling oil can flow all round them.

Fig. 77 (Plate 19) shows a 15,000 kilowatt transformer installed in an outdoor sub-station, with its radiator nearest to us in the picture. This transformer is 'stepping down' from 132,000 to 6,600 volts.

A transformer is very efficient, converting practically all the energy which is put into the primary into energy which is given out by the secondary. In a well-designed transformer the loss is only about 2 per cent.

At this point I can imagine a reader with an enquiring mind (and I am sure anyone who has struggled thus far in my book has got an enquiring mind) saying 'Stop! This is very hard to believe. The voltage of the mains is driving a current through the primary coil of the transformer. The secondary coil is not connected to it in any way whatever. How can it matter at all to the current in the primary what the secondary is doing? To put it in a way which makes it seem most absurd, let us suppose the secondary is cut off from the lamps or motors to which it is giving its energy, so that no current is running in it at all. Surely all the current in the primary is just running to waste.'

This question cannot be answered properly without doing quite a complicated bit of mathematics; engineers, who are not over-fond

PLATE 19

Fig. 77. A transformer in an outdoor substation. Note its cooling radiator in the
foreground (*Metropolitan-Vickers*)

of mathematical equations, sometimes use a graphical method to solve such a problem, i.e. they get the right answer by drawing a diagram. Though it would be out of place to plunge into mathematics here, I am anxious that the reader who has asked a perfectly natural question should have a hint as to the answer.

Suppose in the first place that the secondary circuit is broken at some point, and consider what happens when an alternating voltage tries to drive an alternating current through the primary. At a certain moment, let us say, the voltage is zero, but is mounting up in a direction we will call the positive direction. It starts a current in the primary coil which magnetizes the iron core. Induction now comes into the picture. Because the field inside the coil is increasing, there is a 'back E.M.F.' trying to oppose the increase and so pushing against the voltage. If the voltage keeps on trying to drive the current in the same direction, it will win in the end as it does, for instance, when we make a battery drive a current through an electromagnet. In the present case, however, before the voltage has time to build up a large current, it reverses and tries to drive a current in the opposite direction. Again an induced E.M.F. comes into play and opposes the change, for you will remember the rule that induced E.M.F.s always try to prevent any alteration in a magnetic field. As a consequence, the current through the primary coil is very small, far less than it would be if we applied a corresponding constant voltage to it. We say that the primary coil 'chokes' the alternating current.

When the secondary is connected to a circuit and is giving out energy, a much larger current flows in the primary. It is hard to explain why this is so without the formulae, but perhaps you will see in a general way what is happening. With an 'open'[4] secondary, the current in the primary is choked because it cannot create and

---

[4] An 'open' circuit is one which is broken at some point so that current cannot flow round it.

reverse the magnetic field fast enough. It is like an engine trying to shunt a heavy train backwards and forwards fifty times a second. On the other hand, if a secondary current is able to flow it runs one way when the primary current runs the other way, and they partly cancel each other's magnetic effects. This is what actually happens when the secondary circuit is closed. The primary current is no longer choked to the same extent by its attempts to make and destroy magnetic fields rapidly, because the induced secondary current keeps the fields smaller by always flowing the opposite way. The primary current begins to do more work, and that is where the energy supplied by the secondary comes from.

Here are some figures from the specification of a big transformer, to make things seem more real. The transformer converts 30,000 kilowatts from 132,000 to 6,600 volts or vice versa. When it is idle, the 'magnetizing current' in the primary is only 2.9 amperes, whereas when on full load the current is 131 amperes.

Most of the magnetizing current of an idle transformer, small as it is, does not represent waste of energy. The voltage mounts up and starts a current in the primary. When the voltage dies away to zero, this current goes on running, kept up by the induced E.M.F. It continues to run even when the voltage is reversed, and while it is doing this it *feeds back energy into the mains,* just as a dynamo would. You can imagine it saying to the voltage, 'You've got me started; I'm hanged if I'm going to change just to please you.' In every alternation, the current runs for nearly half the time *against* the voltage, giving back nearly all the energy it gets. The idle transformer therefore absorbs very little energy indeed from the supply to which it is connected.

Currents which are running out of sympathy with the alternating voltage in this way are called 'Wattless amperes.' They do not represent useful energy. When we really want the current to do something, the part of it which is wattless is a nuisance because it only warms up the wires without doing any work. When an engineer says that his 'power factor' is high, he means that most of his amperes are really doing work and not being wattless, or, to put it more precisely, that current and voltage are very nearly in step with each other.

Even if my reader finds this difficult to follow it has perhaps been worth while to hint at these effects if only to show that alternating current behaves in a very different way to direct current, and that we must revise our ideas when we come to deal with it.

5. EDDY CURRENTS; ALTERNATING CURRENT SUPPLY METERS

Meters deserve a section to themselves. I mentioned an alternating current meter at the beginning of the book as an illustration of the difficulty of understanding how electrical machinery works. It really is difficult to understand, even when one has some knowledge of electricity and magnetism. If the reader follows the working of the meter, he may congratulate himself on having got the idea of induction clearly.

We must start off with the idea of *eddy currents*. When a magnet approaches or recedes from a coil of conducting wire, a current is set up in the wire by electromagnetic induction.

Suppose there are two coils of wire A and B, and that a magnet is moved from a position opposite A to one opposite B, as in Fig. 78*a*. What will happen? A current will run in A, which tries to hold the magnet back, and one will run in B, which tries to prevent the magnet approaching — by our law which states that the current always runs in such a direction as to oppose the movement of the magnet. Now suppose that instead of the two coils of wire we have a solid block of copper and move the magnet past it (Fig. 78*b*). Currents will run round in the block of copper as shown by the arrows just as they do round the coils, though they are now not confined to a circular route. An eddy current is set up by the moving magnet in

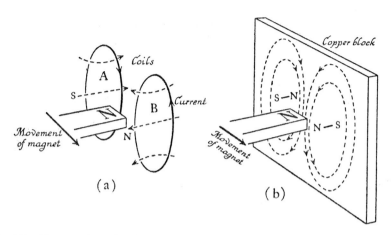

Fig. 78.   The creation of eddy currents in a conductor by a moving magnet.

somewhat analogous fashion to the water eddy around the moving blade of an oar. The eddy current is in such a direction that it opposes the movement of the magnet.

The consequence is that when a magnet is moved past a conductor, such as a sheet of metal, or the conductor is moved past the magnet, there is a reaction between them resembling friction, a kind of sticky drag. Fig. 79 shows a way of illustrating this effect. A copper plate is swinging between the poles of an electromagnet. When the current in the electromagnet is switched on, the plate stops swinging to and fro and slowly oozes back to rest. In Fig. 52*a* (Plate 13) [not included here] of Professor Blackett's large electromagnet you will see a solid aluminium ball falling between the pole pieces. When the magnet is 'on,' the ball drifts like a thistledown, or as if it were falling through thick treacle. In all these cases eddy currents are set up in such a direction that any movement of the conductor is strongly opposed.

The effect was observed by Arago before Faraday's great experiment on electromagnetic induction, and in fact Arago's observation was partly responsible for guiding Faraday in the right direction. Arago found that if a copper disc were placed beneath a compass needle and rapidly rotated, the needle appeared to be dragged

Fig. 79.   The braking of a swinging copper plate by eddy currents. (*'Sphere'*
photograph.)

round by the disc (Fig. 80). This is still the case when a glass plate
is interposed between disc and needle, so that there can be no question of air currents affecting the needle. He was led to the discovery
by noticing that a compass needle in a box with a copper bottom
only made a few oscillations before coming to rest, as if there were
a drag between copper and needle, instead of several hundred
oscillations, as it did when suspended in the open.

A fascinating way of illustrating eddy currents is by means of a
rotating field produced by three-phase current. An iron ring is
wound with an endless coil, and wires are connected to the coil at

Fig. 80. Arago's experiment. A spinning copper disc beneath a sheet of glass drags round the compass needle above the glass.

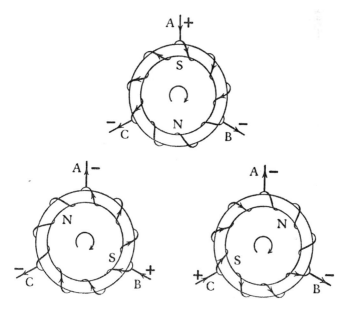

Fig. 81. The creation of a rotating magnetic field by a three-phase current, the three leads being connected to the coil at A, B, C.

three points A, B, C (see Fig. 81). Suppose now that we lead current in at A, and out at B and C. The ring is magnetized with an S pole at A and an N pole opposite A. If the current enters at B, the S pole shifts to B, and so also for C. The coil is now connected at A, B, C to the three terminals of a three-phase supply. You will remember that each terminal of a three-phase supply becomes the positive terminal in turn. The consequence is that the current enters first at A, then at B, then at C. If the alternating frequency is 50 cycles, the north and south poles run round the ring 50 times a second, producing a *rotating magnetic field*.

A shallow bowl of wood or papier mâché is laid on the ring, and a metal ball placed in it. It immediately starts to rotate, and accelerates till it approaches the angular velocity of the field. A metal egg will spin and get up on end. If a copper ring is held in the field one can feel it trying to twist out of one's hand, and directly it is let go it joins the dance. Eddy currents are dragging it round, making it follow the magnetic field.

If you look at the iron core of the armature of a motor or dynamo, you will notice that it is not a solid piece of iron but is built up out of many thin sheets insulated from each other. This is done in order to prevent eddy currents when the armature spins between the poles of the field magnet. If the eddy currents were allowed to flow in the iron they would heat it and fritter away energy.

Fig. 82 shows the way in which eddy currents are used to drive one type of alternating current meter. In Fig. 82*a* (Plate 20) the disc and the magnets which drive it have been removed from the meter and mounted separately. The disc turns freely on a vertical spindle. Now if we were to take a powerful magnet and move it along the surface of the disc near the edge as if stroking it, but without actually touching it, the disc would rotate because of the reaction of the eddy currents. The disc in the meter is being 'stroked' round in a similar way, but instead of using a moving magnet the same effect is produced in an ingenious way by the stationary double magnet seen at the back of the disc. The alternating current supply, passing through the double magnet, sets up a *moving magnetic field* just as if magnetic poles were continually passing over the disc from left to right.

PLATE 20

Fig. 82a.   Disc and electromagnets of an alternating current meter. The
electromagnets are shown enlarged below

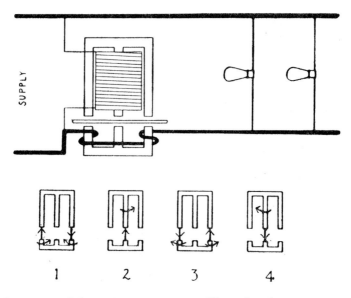

Fig. 82*b*.   Action of alternating current meter. The series of events represented by 1, 2, 3, 4 repeats in each cycle.

Two electromagnets, each with three prongs, are placed opposite each other above and below the disc, as in Fig. 82*b*. The alternating current which is to be measured flows round the coils on the outer arms of the lower magnet, making first one north and the other south, and then reversing the poles, in each alternation. The upper magnet has only got a single coil with many turns on its middle prong. The alternating voltage of the mains across this coil makes this central prong alternately north and south. However, the effect of induction in this coil with its many turns is so strong that the magnetization is always lagging behind the applied voltage. You may remember the argument we used in the case of the trans-former. The voltage tries to magnetize the coil, but is opposed by a back E.M.F. due to induction. The current only gets going just as the applied voltage is dying away and about to change direction. The current goes on even after the applied voltage has changed direction, and is only stopped when the latter has its maximum value in the reverse sense. The current then reverses, and continues in the reverse direction after the E.M.F. has changed back again.

The current behaves, in fact, like the clown in the circus who is trying to help the attendants. He always realizes too late what they are going to do, and rushes to the place to find that they have already done it! The current is 'slow in the uptake' because of the effect of induction.

The magnetic effects may be followed by studying in Fig. 82*b* the series of events 1, 2, 3, 4, which take place during each cycle. In 1 the lower wire from the supply is positive and a current is being driven round the lower electromagnet and through the lamps to the upper wire. In 2 the current in the lower magnet has ceased, but as we have seen the current in the upper magnet is at its greatest. The centre prong is therefore magnetized. Similar events in the reverse direction happen in 3 and 4. The vertical arrows show the direction of the magnetic field in each case. Now if you will follow the arrow pointing upward in 1, which is on the left-hand side, you will note that in 2 it has moved to the middle and in 3 has moved to the right-hand side. The downward arrow starts on the left-hand side in 3, moves to the centre in 4 and then to the right as in 1. The transition from 1 to 2, or 2 to 3, is, however, a gradual one, and not abrupt as in the figure. It is just as if we were stroking the disc with a series of magnets passing from left to right though nothing is actually moving. The stronger the current in the lower magnet, and the higher the voltage driving current through the upper magnet, the greater is the effect on the disc and the faster does it rotate.

The disc is braked by eddy currents, caused by a permanent magnet whose poles are on either side of it (see Fig. 1) [not included here]. When properly designed, the rate of rotation is accurately proportional to the power. The rotations of the disc move the wheels of a counter (Fig. 1) like the cyclometer which counts the revolutions of a bicycle wheel, and this registers the total number of units which have passed through the meter.

Sufficient principles arc involved in the construction of this humble instrument to illustrate a whole text-book of Electricity and Magnetism.

6. POWER STATIONS

A Power Station or Generating Station is a place where mechanical power is generated and turned into electrical power. With very few exceptions, the mechanical power is obtained either by using coal to raise steam or by using the energy of falling water.

Most generating stations in this country, such as the Battersea station shown in Fig. 83 (Plate 21) are steam stations. In former days the steam drove reciprocating engines with cylinders and pistons, but this type has now been superseded by the turbine. A turbine and dynamo are placed end to end with their shafts coupled together, and the combination is called a Turbo-generator.

There is a fine description of the steam turbine in the book *Engines* which Professor Andrade wrote after giving the Christmas lectures at the Royal Institution three years ago, which I warmly recommend you to read if you want to know more about it. I will just remind you that a turbine is a windmill blown round by steam. The rotating part of a turbine, removed from the outer case, is shown in Fig. 84 (Plate 22). The high-pressure steam enters at one end of the turbine and blows against the first set of moving blades set at an angle like the vanes of an iron windmill. It then passes through a set of stationary blades fixed to the outer case, which serve the purpose of directing the steam blast, which has been deflected by the first set of moving blades, back to its former course and on to the next set of moving blades. It passes alternately through moving blades and fixed blades, expanding as its pressure falls, so that the blades have to be made bigger and bigger. Generally there are two turbines, high-pressure and low-pressure.

PLATE 21

Fig. 83.   Battersea Power Station. The ramps on the left convey coal (*Central Electricity Board*)

You can see in Fig. 85 (Plate 22) which shows a turbo-generator in Battersea Power Station, the steam pipes leading from the one to the other. The gale of steam through the turbine, to quote from Professor Andrade's account, may be 75 miles an hour when it enters and 300 miles an hour when it comes out after expansion, taking abont one-sixteenth of a second to go the length of the turbine.

The steam which has done its work passes straight into a condenser, where it turns back into water. It is condensed by coming in contact with pipes through which water is flowing. A large generating station needs about fifteen million gallons of cooling water each hour. It is as important to place a steam station where this water is available as to have a convenient supply of coal. Battersea, for instance, uses the water from the Thames, and Barton uses the Manchester Ship Canal.

The turbine has many advantages as a means of driving a dynamo. It rotates smoothly at a high speed, and as a high speed is also suitable for a dynamo the two can be coupled directly together. It has a high efficiency. A turbine turns about twenty-eight per cent of the energy of the burning coal into useful work, as compared with ten per cent for a first class railway locomotive. This comparison is, of course, not quite fair to the locomotive which cannot carry a condenser. A big power station uses just over one pound of coal to generate a 'unit,' i.e. a kilowatt for an hour. A striking feature of a turbo-generator is its compactness. It is hard to believe that so much power is being developed in so small a space.

A station which derives its energy from falling water is called a hydro-electric station. If the water is coming from a reservoir at a considerable height above the station, so that it is at high pressure, it is used to drive a 'Pelton Wheel.' The water comes out as a narrow jet from a nozzle, and this jet is directed against buckets arranged around the rim of a wheel like a glorified water-wheel. Fig. 86*a* shows a diagram of a Pelton Wheel, and Fig. 86*b* the peculiar shape of each bucket. The object of the nick at the bottom of the bucket seems

PLATE 22

Fig. 84. Blades of a turbine mounted on the shaft (*Metropolitan-Vickers*)

Fig. 85. A turbo-generator in Battersea Power Station (*Central Electricity Board*)

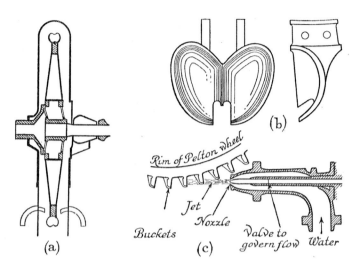

Fig. 86. A Pelton wheel. (*a*) The wheel seen edgewise. (*b*) A 'bucket' with its notched edge. (*c*) The jet hitting the buckets on the rim of the wheel. (*Gibson's Hydraulics.*)

mysterious at first sight. We only want the water to hit each bucket as it comes to the bottom of the wheel, and the nick allows the jet to miss the buckets which have not yet reached the bottom position. Fig. 86*c* shows how the jet plays upon the buckets. If on the other hand the water is only coming from a moderate height, as at Niagara for instance, it is used to drive a turbine. There are many types of turbine and it would take too much space to describe them here. The principle is the same as that of the windmill or steam turbine, but in general the water is led in at the centre of the turbine and shoots over curved blades towards the rim, making the turbine spin round.

Opportunities of using water power mostly occur in sparsely inhabited mountainous country where the power is not needed locally. However, now that electrical transmission of power over long distances is possible, water-power can be used. In our own country, a number of stations in south-west Scotland are being built which will pour their energy into the grid system. Fig. 87 (Plate 23) shows the pipe lines running down hill to a hydro-electric station.

PLATE 23

Fig. 87. The British Aluminium Company's hydro-electric station at Fort William. The photograph has been taken from between the pipes which run down the hill and are seen leading to the station in the distance (*Central Electricity Board*)

To describe the dynamos in any detail would take us too far afield, but I will mention one or two features which may help you to understand what you see if you visit a large power station.

In the first place, if you go back to the diagram of the simple dynamo in Fig. 62 [not included here], you will readily understand how an alternating current is generated. It is simpler to generate alternating than direct current; it is done by leaving out the commutator. Instead of pressing on a divided commutator, the brushes may touch conductors called slip rings, each end of the armature winding having a slip ring and brush of its own. We have seen that the induced current in the armature rushes first one way and then the other, and if we do not rectify this current by a commutator it will be sent out as alternating current. It is more common nowadays, however, to plan the A.C. dynamo in such a way that brushes are unnecessary. Generators work at high voltages, the common ones in this country ranging from 6,600 to 33,000. The difficulty of collecting large currents at these high voltages from rotating armatures is avoided by the cunning device of turning the dynamo inside out. The armature is stationary and the field magnet goes round. The coils in which the current is induced are placed on the inner surface of the outer frame (called the 'stator'), and the field magnet or 'rotor' spins inside. The current exciting the field magnet is relatively small and at a low voltage and is therefore easily led in by slip rings. The powerful high tension currents come from the stator windings, and as these are at rest it is much easier to insulate them. Fig. 88*a* (Plate 24) shows the rotor for one of the 105,000 kilowatt generators at Battersea being despatched on a truck and Fig. 88*b* (Plate 24) shows the casing in which the coils of the stator are placed.

A direct current dynamo can excite its own field magnets, but in an alternating current machine they must be excited by a separate supply of direct current. At the end of a big generator there is a small direct-current dynamo whose sole task is to excite the field magnets of the rotor.

PLATE 24

(a)

(b)

Fig. 88.   The rotor (a) and the stator casing (b) of a large generator
(Metropolitan-Vickers)

The windings of the stator are very complicated, but a simple
model may serve to show how a three-phase current can be
excited. We have already seen (page 168) [p. 257 here] that if a
three-phase supply is fed into three points of a coil round a ring, a

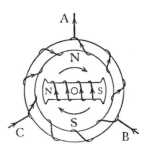

Fig. 89. A (very unpractical) way of generating three-phase current, merely illustrating the principle.

rotating magnetic field is produced. This works backwards. If a magnet is rotated inside the ring, a three-phase voltage is generated. In Fig. 89, suppose a powerful electromagnet is rotating inside the ring. What current will be induced in the coils? It is always safe to give the general answer; it will be a current which *opposes the motion*. As the magnet turns, the induced current will make north and south poles in the ring which rotate ahead of those of the magnet and tend to push it back. We have already seen that a current which produces such a field is a three-phase current. The current will come from the leads in the order A, B, C, etc. We have, in fact, designed a crude three-phase dynamo, with the coil as stator and the central electromagnet as rotor.

When a number of generators are feeding alternating current to the same mains, they all automatically keep step like soldiers, timing their impulses of current so that they occur simultaneously. If one generator falls a bit behind in its timing, the current from the others tends to speed it up as if it were a motor. If it gets a bit ahead, it does more than its share of the work till it falls into line again. When it is required to add an additional generator to those already working, it is speeded up till an instrument with a big dial informs the operator that it is keeping time with the running machines. It is then switched on to the 'bus-bars,' or common terminals of the main supply, and settles down into the collar along with the rest of the team. Hitting off the right moment to close the switches is rather like making a quiet gear-change in a car.

7. THE LOAD ON A POWER STATION

The amount of electrical power which is used varies considerably during the day, and it is different in summer and winter. To illustrate this, I have obtained from the Barton Power Station in Manchester graphs showing the output on two typical days, 5th July and 16th November, 1934. The hours run from left to right on the diagram, starting from midnight. The output is measured in mega-watts, a mega-watt being 1,000 kilowatts or about 1,300 horse-power. Taking the summer day first (Fig. 90a) you will see that from midnight to 7 a.m. little current is used. Between 7 and 8 there is a sharp rise because people go to work and machines in factories are set in motion. The demand is continuous during the morning till the factory hooters sound at 12 mid-day. The machines stop, and everyone hurries off to lunch. At one o'clock the output rises again as work is resumed, but it may perhaps be significant that the rise is more gradual than the sudden drop at 12! Between 5 and 6 in the afternoon work ceases and people go home. As it is nearly midsummer, they do not need to light up their houses, and the current remains low till nearly ten o'clock when there is a little peak after sunset. Manchester is a sober and industrious city, and not a gay metropolis like London, so you will notice that soon after this everyone goes to bed.

In winter (Fig. 90b) the current is much the same in the small hours, but is considerably higher after 8 a.m. The winter gloom makes it necessary to use a good deal of electric light in factories, shops, offices, and homes during the day. You will see the same fall at lunch-time. At half-past three it becomes dark, and all the lights go on, giving a tremendous peak. This is the heaviest load the station has to cope with. It remains high after work is finished, because

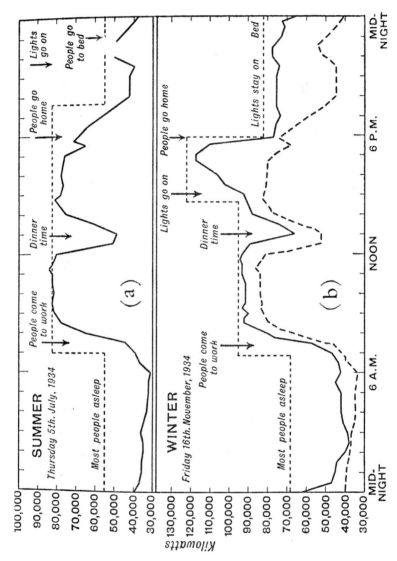

Fig. 90.   The fluctuating demand for electrical power, during typical 24-hour periods in summer and winter. (*Barton Power Station, Manchester.*)

extra lights go on in the houses, and it does not drop till people go to bed and the street lights are reduced between 11 and 12. The records of the station are like a little history of everyone's habits.

The upper dotted line shows how the load is sustained. The power station has several big turbo-generators, but only as many are set going as are required to meet the demand for power. A step up in the dotted line shows the increase in possible output when a new generator is set in motion; the dotted line must always be above the line which indicates the demand. When an additional generator is working more steam is required, so demands must be anticipated as far as possible. The demand depends on such factors as whether it is fine or overcast, hot or cold, as well as on the time of day and the season. These graphs represent the situation when each station had to meet the fluctuations in its own local demand. Now that stations are linked by the grid, there is a scheme of central control described in the next section.

8. THE GRID

The grid is a system of high voltage transmission lines all over the country which is run by a body under public control, the Central Electricity Board. The system illustrates in a very fascinating way how the problem of providing a general supply of electrical power at cheap rates may be attacked. A map of the grid is shown in Fig. 91.

To understand the purpose of the grid, we must realize that there are two stages in electrical supply, which may be compared roughly to manufacturing and selling in other trades. There is in the first place the generation of the power and its transmission to distributing centres, or the wholesale supply of power. In the second place, lines must be laid and maintained, from these distributing centres to customers, and the current used by customers must be measured and charged for, this being the retail side. As in many other trades, the cost of distribution is immensely greater than that

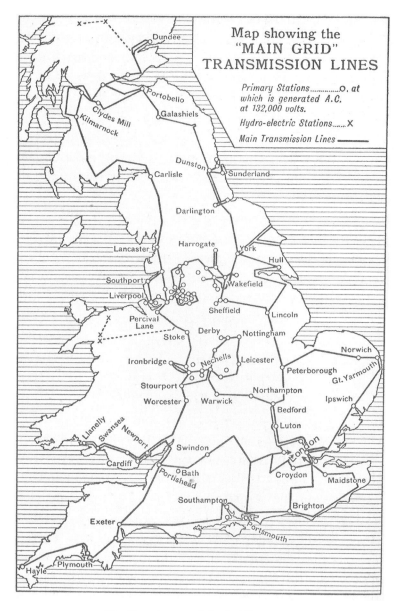

Fig. 91. A map of the 'grid' system.

of production. Electrical power in bulk is astonishingly cheap; the tariff fixed by the Central Electricity Board varies somewhat from one area to another, but in the North-western area for example, the running charge is one fifth of a penny for each unit of a kilowatt-hour.[5] When we pay the quarterly bill for 'Electricity' a very small fraction of the charge represents the cost of the electrical power we have used. What we are really paying for is the cables, wiring, meters, and services which make the power available to us. Domestic tariffs vary in different parts of the country, but perhaps I may take that in my own neighbourhood as an example. In every £10 of my bill, just over £1 represents the cost of producing the electrical power I use. The remaining £9 is paid for the privilege of being able to switch on the light when I like. As consumers, we are like people who have hired a box at the opera for the whole season and only visit it two or three times, or who keep a taxi waiting all day in case they should need it. If we all used our supply to its full extent continuously, the cost per unit would be extremely small. It costs much more because we only use it at odd times, and, worse still, we all want it at the same time, producing the big peaks in the load which have to be coped with by the supply stations.

Before the grid was built, many local supply undertakings both produced their power and sold it to their customers. Each station had to have a sufficient reserve of machinery to meet its peak load. Though large stations produced power cheaply, many of the small stations were very inefficient. Some produced direct current, some alternating, voltages varied, and frequencies varied, so that customers had to buy lamps and motors to suit their particular local supply.

The change made by the grid scheme is briefly as follows: Each local undertaking still has control of the lines to its customers. It

---

[5] This rate is practically doubled by a fixed charge made by the Board to meet the interest on their capital expenditure on plant.

fixes retail tariffs and collects payment. On the other hand, responsibility for the production of electrical power has been taken over by the Central Electricity Board. Though the big supply stations still run their generators, they are under the control of the Board, which has in effect bought up the manufacturing side, and sells energy again to the retailers to distribute to their customers.

In order that it should be possible to pool the electrical supply in this way, each generating station is linked to the grid by transformers. We can think of it most simply by imagining the generators pouring their energy into the grid, under the control of the Board, and supply undertakings drawing energy from the grid and paying the Board for it. Actually it seems a bit more complicated in many cases because the generating stations feed local customers' lines direct without passing the energy into the grid and then drawing it back again, but this is only a matter of book-keeping. If the local generators are not working at all, the supply company draws all its energy from the grid through the transformer. If the generator is working full steam ahead, and the local demand is light, it is pouring its extra energy into the grid through the same transformers.

What is the object of this? It is done in order to meet the demand in the most efficient way. Formerly each station had to keep its machines running the whole time, raising more steam when the demand was heavy and slacking off when it got light. Now the stations are marshalled in groups, like soldiers being brought into action under the orders of a general. The huge stations work day and night at full power, meeting what is called the 'Base Load.' The next group of stations, called 'Two-Shift Stations,' are shut down at night and at the week-end, when the demand is light. Finally there is a group of reinforcements called Seasonal Stations, which are only called up during peak-load time in winter (see Fig. 90) and are shut down all the summer. The most efficient stations work the whole time, and the least efficient only for a short time. Many of the smaller undertakings have shut down their generators altogether and just act as retailers, buying power from the grid and selling it to their customers. It is much easier for a new supply company to start, because it need not build a generating station. It can link on to the grid and set up shop.

The way in which this pooling of power is carried out is very fascinating. There are two sides to it, the central control and the grid links between stations.

The country is divided into areas, each with its own network of lines to which all stations in that area are linked, and with a control centre which is like the brain of the organization. If one goes into a control centre one sees a vast array of dials on panels in front of the engineer in charge (Fig. 92, Plate 25). These dials are labelled with the names of generating stations and are recording what that station is sending into the grid or taking from it at that moment, although the station may be a hundred miles away. There is also a big diagram of the grid to show how the stations are linked up, with red lights to indicate when switches are on and green lights when they are off. One can see exactly what is happening over the whole area.

The control station is able to forecast with fair accuracy what the demand will be at different times of the day, for the particular times of the year. It issues orders beforehand to generating stations, telling each during what hours it is to run and what energy it is to produce. The extra bit, which cannot be foreseen, is taken on by one big station. All the other stations are going full steam ahead, but at this particular station the engineer has (figuratively) his hand on the throttle. If the demand goes up, he lets more steam into his turbines, if it goes down he shuts some off. Should the demand rise so quickly that he cannot cope with it, he telephones the control centre which calls up reinforcements by telling a new station to get going.

How does the man in control know when to put extra power into the grid and when to slack off? The way in which this is done reminds us of a runner or cyclist acting as pace-maker to a group. He does it by watching the *'frequency' of the system*. The normal

PLATE 25

Fig. 92. The room at a control centre where the supply of power to the Grid is regulated (*Central Electricity Board*)

frequency is 50 cycles, and the generators in all the stations are turning round in step with each other, for we have seen that this keeping in step is a consequence of alternating current supply. If a number of people switch lights and motors on, *all* the generators begin to run more slowly, still keeping time. The needle of the frequency meter, which you see at the centre of the control room, drops below 50, and the generators at the pace-making stations are boosted up to get it back to 50 again. It works the other way, of course, when the demand falls.

Underneath the frequency meter is a clock with two hands. One hand is run by a chronometer which is checked against Greenwich time. The other hand is run by an alternating current motor, much as in the electric clocks which are now common (Fig. 93). If the frequency is exactly 50, the two hands will keep in step. If the electric hand falls behind the chronometer hand it shows that the generators

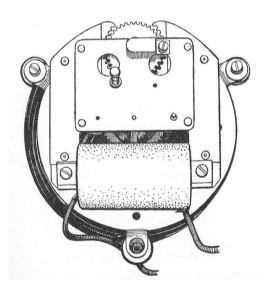

Fig. 93.   An electric clock. The toothed iron wheel seen above moves on one tooth each time the alternating current excites the electromagnet with toothed poles. The wheel drives the clock hands through a chain of gears. (*Ferranti.*)

are turning too slowly, and the grid frequency is boosted till the electric hand catches up, and vice versa. At any one moment there may be a little difference between the two hands but this is never allowed to exceed a few seconds. That is why our electric clocks run by the alternating current supply keep such excellent time. As long as they do not stop, they *cannot* be more than a few seconds wrong. Their motors are rotating in step with the generators in the power stations, and are being forced to keep time by the whole grid system backed up by the authority of the Astronomer Royal at Greenwich.

9. TRANSFORMING AND SWITCHING SUB-STATIONS

The enclosures one sees in many places with a network of girders and wires on insulators, and rows of iron tanks beneath with porcelain horns like the antennae of immense insects, are *Transforming and Switching Sub-stations.* There is generally one outside each generating station, linking the station to the grid. Many of the sub-stations, however, are not attached to a generating station. They are merely places where energy is drawn from the grid and sent out again at a lower voltage for local supply.

A typical sub-station of this latter kind (Little Barford) is shown in Fig. 94a, Plate 26. It looks an extremely complicated maze, but actually it is doing a very simple job. A main transmission at 132,000 volts linking Peterboro' and Bedford runs through the sub-station. Two transformers draw energy from the main lines and transform down to 33,000 volts for local lines, which lead to various places in the neighbourhood. In the photograph the lines at 132,000 volts are seen on a tower in the foreground, and the local supply is led away from the distant end of the enclosure.

Why is it so complicated, when one would think it only necessary to join the transformer to the cables? There is so much gear because of the care which has to be taken in handling large

PLATE 26

Fig. 94a. Little Barford Transforming and Switching Sub-station (*Central Electricity Board*)

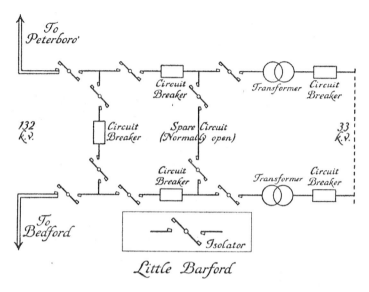

Fig. 94 [sic; Fig. 94*b*].   A plan of the sub-station shown on the opposite page. (*Central Electricity Board.*)

amounts of power at high voltage. I have put a plan of the station on the opposite page (Fig. 94*b*) representing the three lines required for three-phase supply as a single line throughout in order to make the diagram easier to follow. The 'isolators' are arms which swing round, joining lines together when they are closed and disconnecting them when open. They are only used to disconnect the lines when no current is flowing. The 'circuit breakers' on the other hand are switches which can be used to stop the current and which also are thrown open automatically if the current rises to a dangerously high value. They are essential for protection. Sometimes there is a 'flash-over' on the transmission lines. One cause of this is fog, which deposits a conducting film of moisture on the insulators so that a discharge starts between the cables followed by a great blaze which effectively short-circuits them. This would spell disaster to generators and transformers if they were not protected by the circuit breakers, which immediately 'trip' and cut off the current. In Fig. 94*b*, all the insulators are closed in normal working except those on the spare circuit. If anything happens to the Peterboro' line the two circuit breakers nearest to it immediately break contact. One of the transformers still gets energy from the Bedford line, so the local supply is not cut off. If a short circuit occurs in the local supply, the circuit breakers between it and the transformers are tripped. In that case all the lights in the neighbourhood go out (you will know how this sometimes happens) but the transformers and grid are protected, and the supply will come on again directly the trouble is put right. Everything is safeguarded as far as possible.

Any part of the system which has to be put right is of course first cut off by the isolators so as to make it safe to approach it. If you realize the need for all this gear, and for the wide separation and insulation of the high tension wires from each other, you will see why even a simple sub-station like this is so large.

The circuit breakers are immense affairs. At first sight it would seem that a switch is the simplest thing possible. If two pieces of metal touch, the current flows; if they are separated it cannot flow. However, even in our house supply a switch has quite a complicated mechanism (Fig. 95). If you unscrew the cover, and work the light switch, you will see that when it is turned off the contact comes out with a snap, being jerked out by a powerful spring. If the contacts were separated slowly, there would be an arc which might melt the metalwork of the switch. If such precautions have to be taken for 230 volts, you can imagine what an engineering problem a switch for 132,000 volts presents. Contact is made by metal plungers which fit into holes. An electro-magnet beside the switch compresses a powerful spring, and when the switch is tripped the spring pulls the contacts out with terrific force. Several of the circuit breakers can be seen in Fig. 94, Plate 26. The cylindrical tanks contain the oil in which the switches work, and the wires lead into the tank through long porcelain horns so as to be safely insulated.

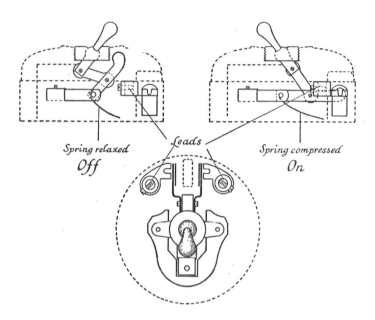

Spring relaxed   Leads   Spring compressed

Off   *On*

Fig. 95.   A domestic switch.

The fuses in a house supply have the same protective duty on a small scale as the circuit breakers. The ordinary fuse is a piece of wire either of an alloy which melts easily or of thin copper. There is generally a pair of fuses in the fuse-box for each group of three or four rooms in the house. If the wires to the lamps are allowed to touch so that there is a short, so large a current runs through the fuse that the wire melts, and any damage from the excessive current is prevented.

The towers which carry the transmission lines (Fig. 96, Plate 27) are so well known that I will only draw your attention to the *three* conductors (or six on a double circuit) for the three-phase current, the earthing conductor attached to the top of the tower, and the strings of insulators. The conductors are about $\frac{3}{4}$ inch in diameter, and are made of strands of aluminium to give a high conductivity with a core of steel strands for strength.

PLATE 27

Fig. 96. Towers carrying 132,000 volt transmission lines of the Grid (*Central Electricity Board*)

Descriptions such as these are very dull affairs unless one also looks at the actual machinery. However, you will find a visit to a power station far more interesting if you understand the way it works, and I hope that the account I have given here will enable you to talk as a brother expert to the engineer who is taking you round, and to plunge into technical discussions with him as to how such vast amounts of power are safely handled.

Objects and Pictures

Richard Langton Gregory*

We are surrounded by objects. Our lives are spent identifying, classifying, using and judging objects. Objects are tools, shelter, weapons; they are food; they are things precious, beautiful, boring, frightening, lovable . . . almost everything we know. We are so used to objects, to seeing them wherever we look, that it is quite difficult to realise that they present any problem. But objects have their existence largely unknown to the senses. We sense them as fleeting visual shapes, occasional knocks against the hand, whiffs of smell — sometimes stabs of pain leaving a bruise-record of a too-close encounter. What we experience is only a small part of what matters about objects. What matters is their 'physical properties', which allow bridges to stay up, and car engines to run, though the insides are hidden. The extraordinary thing is how much we rely on properties of objects which we seldom or never test by sensory experience.

It has sometimes been thought that behaviour is controlled by information immediately available to the eyes and other senses. But sensory information is so incomplete — is it adequate to guide us among surrounding objects? Does it convey all that we need to

* The original version appeared in: Richard Langton Gregory, Chapter 1, *The Intelligent Eye*, London, 1970, pp. 11–31.

know about an object in order to behave to it appropriately? At once we see the difficulty — the continuous problem the brain has to solve. Given the slenderest clues to the nature of surrounding objects we identify them and act not so much according to what is directly sensed, *but to what is believed*. We do not lay a book on a 'dark brown patch' — we lay it on a table. To belief, the table is far more than the dark brown patch sensed with the eyes; or the knock with the knuckle, on its edge. The brown patch goes when we turn away; but we accept that the table and the book remain.

Bishop Berkeley (1685–1753) questioned whether objects in fact continue to exist when not sensed — for what evidence could there be? But rather than allow objects to have, as Bertrand Russell puts it, 'a jerky life', he supposed that they exist continuously because God is always observing them, which Berkeley used as an argument for the existence of God. His doubt, and later certainty, are expressed in Ronald Knox's famous limerick and reply:

There was a young man who said, 'God
Must think it exceedingly odd
    If He finds that this tree
    continues to be
When there's no one about in the Quad.'

Dear Sir:
    Your astonishment's odd:
I am always about in the Quad.
    And that's why the tree
    Will continue to be,
Since observed by
    Yours faithfully GOD

Berkeley's doubt raises an important question: what can we *know* beyond sensation?

The optical images in the eyes are but patterns of light: unimportant until used to read non-optical aspects of things. One cannot eat an image, or be eaten by one — in themselves images are biologically trivial. The same is not, however, true for all sensory information. The senses of touch and taste do signal directly important information: that a neighbouring object is hard or hot, food or poison. These senses monitor characteristics immediately important for survival: important no matter what the object may be. Their information is useful before objects are identified. Whether the hand is burned by a match, a soldering-iron or boiling water makes little difference — it is rapidly withdrawn in any case. What matters is the burning heat, and this is directly monitored. The nature of the object may be established afterwards. Such responses are primitive — pre-perceptual reactions, not to objects but to physical conditions. Recognising objects, and behaving appropriately to their hidden aspects, comes later.

In the evolution of life the first senses must have been those which monitor physical conditions which are immediately important for survival. Touch, taste and temperature senses must have developed before eyes: for visual patterns are only important when interpreted in terms of the world of objects. But this requires an elaborate nervous system (indeed almost a metaphysics) if behaviour is controlled by belief in what the object is rather than directly by sensory input.

A curious hen-and-egg type of question arises: which came first, the eye or the brain? For what use is an eye without a brain capable of using visual information — but then why should a 'visual' brain develop before there were eyes to feed it with visual information?

What may have happened is that the primitive touch nervous system was taken over to serve the first eyes, the skin being sensitive not to touch only but also to light. The visual sense probably developed from a response to moving shadows on the surface of the skin — which would have given warning of near-by danger — to recognition of patterns when eyes developed optical systems. The stages seem to have been, first a concentration of specially light-sensitive cells localised at certain regions, and then 'eye pits', the light-sensitive cells forming the bottom of gradually deepening pits which served to increase the contrast of shadows at the light-sensitive regions, by shielding them from ambient light. The lens most probably started as a transparent window, protecting the eye pits from being blocked by small particles floating in the sea in which the creatures lived. The protective windows may have gradually thickened in their centres, for this would at first increase the intensity of light on the sensitive cells until — dramatically — the central thickening produced an image-forming eye: to present optical patterns to the ancient touch nervous system.

Touch can be signalled in two quite distinct ways. When an object is in contact with an area of skin, its shape is signalled from many touch receptors, down many parallel nerve fibres simultaneously to the central nervous system. But shape can also be signalled with a single moving finger, or probe, exploring shapes by tracing them in time. A moving probe can not merely signal the two-dimensional shape that happens to be in contact, but can trace shapes in three dimensions, though it will take a considerable time to do so. Also, if the object it is exploring is itself alive, it will certainly give the game away — as we know by being tickled.

Touch is not a secret sense, and it is limited to objects in physical contact. This means that when a foe is identified by touch, it is too late to devise and carry out a strategy. Immediate action is demanded, and this cannot be subtle or planned. Eyes give warning of the future, by signalling distant objects. It seems very likely that brains as we know them could not have developed without senses — particularly eyes — capable of providing advance information, by signalling the presence of distant objects. As we shall see, eyes require intelligence to identify and locate objects in space, but intelligent brains could hardly have developed without eyes. It is not too much to say that eyes freed the nervous system from the tyranny of reflexes, leading to strategic planned behaviour and ultimately to abstract thinking. We are still dominated by visual concepts. Our problem now is to understand the world of objects without being limited by what we have learned through the senses.

The data that most philosophers consider are limited to sensory experience. This is not so for physics, which accepts data from instruments capable of monitoring characteristics of the world quite unknown before instruments were invented. Radio and X-rays were totally unknown to brains until less than a century ago: they have changed our intellectual view of the world, though not sensed directly. This presents something of a paradox for empiricist philosophy, for science uses 'observational data' which can only be 'observed' with instruments: so the senses can no longer be said to be the sole source of direct knowledge.

Since perception is a matter of reading non-sensed characteristics of objects from available sensory data, it is difficult to hold that our perceptual beliefs — our basic knowledge of objects — is free of theoretical contamination. We not only believe what we see: to some extent we see what we believe.

A central problem of visual perception is how the brain interprets the patterns of the eye in terms of external objects. In this sense 'patterns' are very different from 'objects'. By a pattern we mean some set of inputs, in space or time, at the receptor. This is used to indicate and identify external objects giving rise to the sensory pattern. But what we perceive is far more than patterns — we perceive *objects* as existing in their space and time.

An initial problem is how objects are distinguished from their surroundings. This problem becomes clear if we look at a picture where the object is difficult to distinguish. Figure 1 is a photograph of a spotted dog against a dappled background — it is quite difficult to see the dog. Contours and differences of texture or colour help,

Fig. 1.   Retinal images are patterns in the eye — patterns made up of light and dark shapes and areas of colour — but we do not see patterns, we see objects. We read from pictures in the eye the presence of external objects: how this is achieved is the problem of perception. Objects appear separate, distinct; and yet as pictures on the retina they may have no clear boundaries. In this photograph of a spotted dog, most half-tones have been lost (as in vision by moonlight) and yet we can distinguish the spots making up the dog from similar spots of the background. To make this possible there must be stored information in the brain, of dogs and thousands of other objects.

but quite often boundaries of objects are not sharp and colour differences can be misleading. There is a similar problem in hearing speech or music. Words sound distinct from each other, but physically they are not separated. Physically they run into each other, just as the images of objects do upon the retina. Objects are somehow extracted from the continuous patterns at the receptors.

There is a well-known visual effect: 'figure-ground reversal'. Figure 2 shows a face — or does it? Here perception fluctuates between two possibilities. This is important, for it shows at once that perception is not simply determined by the patterns of excitation at the retina. There must be subtle processes of interpretation, even at this elementary level.

The psychologist whose name is associated with figure-ground reversal is the Dane, Edgar Rubin. He used simple but cunningly contrived line drawings in which a pair of shapes, either of which

Fig. 2.   This is seen sometimes as a face, sometimes as something else. Perception fluctuates between two clearly defined possibilities. This is an example of visual 'reversal', by the Danish psychologist, Edgar Rubin.

taken alone would be seen as an object of some kind, share a common border-line. What happens is that, when joined, each competes with the other. Alternatively, one is relegated to mere background, and hardly seen, while the other dominates as object: then this one fades perceptually away to become for a time mere background in its turn. This spontaneous alternation of figure and ground is a curious effect. It shows something of the dynamic nature of perceptual processes.

There are many subtle effects related to figure-ground reversal. When a region becomes figure, it generally looks quite different. Of figure 3 Rubin says:

One can experience alternately a radially marked or concentrically marked cross. If the concentric cross is seen as figure after the radial one, it is possible to note a characteristic change in the concentric markings which depends on whether they belong to the figure or the ground. When they are part of the ground, they do not appear interrupted. On the contrary, one has the impression that the concentric circles continue behind the figure. Nothing of this kind is noticed when the concentrically marked sectors compose that which is seen as figure.

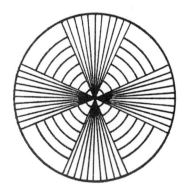

Fig. 3. One of Rubin's figure/ground reversing figures. Here, there are two equally held figures, which in turn are relegated to 'ground'. Regions accepted as 'figure' are subtly changed perceptually, as described by Rubin in the quotation.

Rubin was well aware of the significance of his demonstration-experiments to the problem of how we see objects, though curiously this aspect of his work was largely ignored by later writers. Rubin says in this connection:

When a reversal of figure and ground occurs, one can observe that the area affected by the shape-giving function of the contour at the same time obtains a characteristic which is similar to that which leads one to call objects 'things'. . . . Even when the figure does not look like any known thing, it can still have this thing-character. By 'thing-character' we mean a similarity to what is common to all experienced objects . . .

So we have a hint as to how we might discover the kinds of features used by the brain to make this decision: what are objects, and what is space between the objects? This hint has still not been fully followed up.

Rubin's most striking example is given in figure 4. This is seen alternately as a pair of faces 'looking at each other' and as a vase, which becomes the space between the faces when they are object. Seeing the vase perceptually fade away, to be replaced by the pair of faces emerging from sinister shadows is a queer, almost frightening, experience. Rubin says of this figure:

The reader has the opportunity not only to convince himself that the ground is perceived as shapeless but also to see that a meaning read into a field when it is figure is not read in when the field is seen as ground.

He considers some implications to art, especially the emotional significance of pictures. Again the significance is given not by the stimulus pattern directly but by the interpretation put upon it. As he says charmingly:

If a figure looks like a beloved and admired professor from his homeland, this may remind the subject of the pleasure in having met him again as he

Fig. 4. Rubin's most striking example of visual reversal. This is seen, alternatively, as a pair of faces 'looking at each other' and as a vase.

Fig. 5. The same reversal occurs whether the faces are black and the space between — the vase — is white, or whether the faces are white and the vase black.

stopped by on the way to Göttingen. If a figure looks like a beautiful female torso, this indubitably calls forth certain feelings.

Extending this to works of art:

When one succeeds in experiencing as figure areas which are intended as ground, one can sometimes see that they constitute aesthetically displeasing forms. If one has the misfortune in pictures of the Sistine Madonna to see the background as figure, one will see a remarkable lobster claw grasping Saint Barbara, and another odd pincer-like instrument seizing the holy sexton.

Unfortunately we still do not know in detail which are the features which *prevent* figure-ground (or 'object-space') reversal. These are, however, important. Small areas enclosed in larger areas are taken as figure, or object. Repeated pattern is taken as belonging either to figure or to ground but not to both. Straight lines are attributed to figure. Emotionally-toned shapes are also attributed to figure and, when present, tend to make figure dominant. In addition, the observer's perceptual 'set' and his individual interests can bias the situation.

Rubin used line drawings almost exclusively for his perceptual experiments, as did almost all psychologists until recently. But as we will try to show, pictures are in some ways highly artificial inputs for the eye. Although we can learn a lot about perception from pictures, and they are certainly convenient for providing stimulus patterns, they are a very special kind of object which can give quite atypical results. Object-space reversals can take place (for example when we look at roofs of houses against an evening sky) as well as the figure-ground reversals described by Rubin for pictures.

Object-space reversals merit further study, for what happens as we gradually introduce more data showing that a certain shape *is* an object?

How are some patterns established as representing objects? The problem is acute, for we often see patterns without attributing 'thingness' to them. We see patterns of leaves, of clouds, of fine or coarse texture on the ground. The decorative arts present formal or random patterns, which we may see as patterns not as objects. True, we may almost 'see' Queen Victoria in a cloud formation, or a wicked face, fleetingly, in the flickering flames of a fire. We may see as it were *hints* of objects in patterns, and random shapes, but we certainly can see patterns without accepting them as objects.

The Gestalt psychologists, in the early part of the century, made much play of 'perceptual organisation': that there are Principles, largely inherited, by which stimulus patterns are organised into 'wholes' (*Gestalten*). This organising into 'wholes' was demonstrated with black and white figures, mainly patterns composed of dots. The point is that even an array of random dots tends to form 'configurations'. It is almost impossible to see three dots, with any spacing, without also seeing at the same time a triangle. In a random array we see triangles, squares, rows — all sorts of figures emerge. Patterns of dots were used to try to establish Laws of Organisation. These were discussed in a classical paper by Max Wertheimer, entitled: 'Principles of Perceptual Organisation' (1923). Wertheimer presented several patterns, such as figure 6, and pointed out that they are seen as groups of dots, the closer dots forming perceptual pairs. They are seen as 'belonging' to each other.

Another example given was figure 7. Each sloping line forms a 'unit'. They may also combine to form sloping rectangles. This shows that *proximity* is a factor in perceptual grouping.

Another Principle is shown in figure 8. The circles and the squares are seen separately, each forming rows. This demonstrates that *similarity* is a factor in perceptual grouping. The Gestalt writers

Fig. 6. Simple dot patterns were used by the Gestalt psychologists to investigate their 'Principles of perceptual organization'. We may think of these as primary stages in perception — the linking of data from retinal patterns in terms of probable objects. This may be regarded as like a detective gathering and combining available clues used for making decisions on who is the criminal — or in perception, what is the *object*? This figure illustrates how points close together are seen to 'belong' to each other: the pattern is seen as *pairs* of dots.

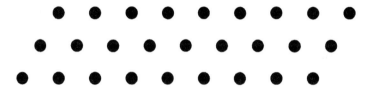

Fig. 7.   Another example of grouping of dots by *proximity*. This figure is generally seen as sloping triads of dots, also as oblique rectangles.

put a lot of weight on what they called 'good figure', and 'closure'; by this they meant properties such as geometrical simplicity, particularly approximations of circles, tending to 'organise the parts into wholes of these shapes'. Considering movement, 'common fate' was regarded as important — that related movement of parts makes the parts cohere into a 'whole'.

To the Gestalt writers, these Organising Principles were innate, inherited. They gave very little weight to individual past experience. Clear evidence of perceptual learning was perhaps lacking at that time and there were reasons, stemming from the contemporary German metaphysics, which made emphasis on innateness attractive. But it is perfectly possible to accept their observations as valid while denying that they are due to innate organising principles. There is no strong argument against saying that most objects are rather simple, and closed in shape, that the parts of objects

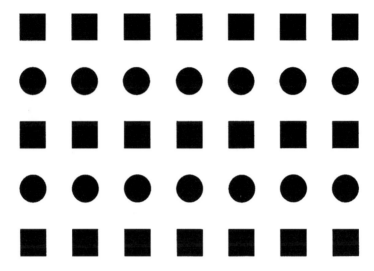

Fig. 8.   This is seen as a series of rows, the identical symbols being perceptually linked. This illustrates the importance of *similarity* in perceptual organisation.

move together, that objects often have repeated structure or texture patterns — and so on. In short, there seems no evidence against supposing that the organising principles represent attempts to make objects out of patterns — typical object characteristics being favoured — and the 'Principles' were developed by inductive generalisation from instances. Since we all experience essentially similar objects it would not be too surprising if we developed similar, even identical, generalisations. Although this was considered by some of the Gestalt writers it was rejected, though not for reasons that we would now accept as having any force. But possibly they were right, possibly there are innate tendencies to organising parts into wholes. This would still be accepted by some writers, though the general Gestalt philosophy is very largely rejected as being an extreme case of non-explanation.

It is probably fair to say that the Gestalt writers were more interested in how we see patterns than in how we see objects, for

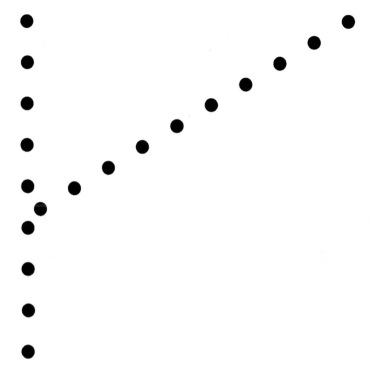

Fig. 9. This shows that there is more to perceptual 'grouping' than proximity. The lowest dot of the oblique row is closest to the dots in the vertical row. Evidently the tendency to organise dots into rows is stronger than the tendency of association by proximity.

they generally used highly artificial visual material, such as the dot patterns.

Wertheimer claims that:

Perceptual organisation occurs from above to below; the way in which parts are seen, in which the sub-wholes emerge, in which grouping occurs, is not an arbitrary, piecemeal and-summation of elements, but is a process in which characteristics of the whole play a major determining role.

But if this were true in normal perception, we might expect that the world would look like a wobbly jelly.

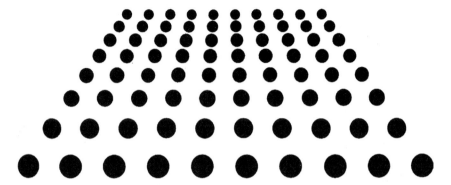

Fig. 10.   Lines of dots (or solid lines) which converge, are perceptually organised in three dimensions. Normal objects exist and are seen in three-dimensional space, though pictures — including retinal images — are flat. Converging features are generally taken to indicate depth by perspective shrinking of the image with increasing distance. For normal retinal images this is a good bet, but in pictures the convergence of perspective is presented to the eye on the flat plane of the picture — making pictures essentially paradoxical, Here the organisation is appropriate to normal three-dimensional space and not to the converging dots which lie on the flat plane of the page. They are organised in terms of normal space rather than the picture-plane.

Granted that organising into 'wholes' is important, it is easy to find examples of where this must be due to individual past experience. For example, we find it in grouping letters in a language — which is most certainly learned. There used to be an English comedian, who called himself NOSMO KING. This was not in fact his real name. He 'discovered' it one day seeing, written across a double door at the theatre: NO SMOKING. When the doors opened he saw his new name appear.

This was anticipated by Helmholtz, the great German physicist (1821–94), when he argued for individual learning as important for perception, in the following:

There are numerous illustrations of fixed and inevitable associations of ideas due to frequent repetition, and even when they have no natural

Fig. 11. This oblique camera angle of a textured surface gives a compelling impression of a slanting surface. The bricks are seen as slanting, yet the page on which the picture lies is not — and does not appear to be — slanting. This double reality is part of the paradox of pictures.

*connections, but are dependent merely on some conventional arrange-
ment, as, for example, the connection between written letters in a word
and its sound and meaning. . . . Facts like these show the widespread
influence that experience, training and habit have on our perceptions.*

Few would disagree with him, but there remains the possibility
that some perceptual organising processes are 'wired in' at birth.
For reasons of economy, we might expect it. At any rate it is now
clear from neurological studies that some visual feature-detectors
are built in to retina and brain structure. This has been established
over the last ten years from direct electrical recording from individual nerve cells of the eye and the brain.

When the electrical activity of local regions of a frog's retina are
recorded from electrodes placed in excised eyes, it is found that only
a few features of the pattern of stimulation on the retina are represented in terms of neural activity, and so signalled to the brain.
Signals are given when the stimulus light is changed in intensity —
some cells signalling when it is switched on, others when it is
switched off, while others signal any change. (They are known as
'on', 'off' and 'on-off' receptors.) Receptors signalling changes of illumination are probably responsible for signalling movement, which is
vital for the frog for detecting and catching flies; as indeed it is to all
animals for survival, since movement is generally associated with
potential food or danger. At the first stage of visual perception — the
retina — we find neural mechanisms responding to specific patterns
in space or time. In a delightful paper, 'What the Frog's Eye Tells the
Frog's Brain', by Lettvin, Maturana, McCulloch and Pitts, several specific pattern-receptive mechanisms are identified. The eye responds
to movement, to changes of illumination and to what we may call
'rotundity'. A small black shadow is signalled strongly and serves to
evoke the fly-catching reflex. This 'bug-detector' gives an immediate
response — the tongue shooting out for fly-catching — without loss
of time by information processing by the brain. The frog's brain

receives but few kinds of pattern information from its eyes. In general as brains develop up the evolutionary scale eyes signal more information and become simpler. The retina is not merely a layer of light-sensitive cells, it is also a 'satellite computer' in which visual information is pre-processed for the brain. Vital information, such as movement, is extracted and signalled directly by the retina in eyes as highly developed as the rabbit and, most likely, in our own.

A really basic discovery was made by two American neurophysiologists, D. Hubel and T.N. Wiesel, some ten years ago. Placing micro-electrodes in the 'visual brain' (the *area striata* of the occipital cortex) of the cat, they found that certain brain cells respond to specific patterns at the eye, other brain cells to other patterns. Some cells responded to movement in one direction but not the opposite or any very different direction; other cells responded to lines oriented at a certain angle; others to corners. Activity at the surface of this region of cortex corresponds rather crudely to the spatial position of stimulation of the retina — a rough electrical spatial map is projected on the occipital cortex from the eye — but deeper down spatial position is lost and patterns are represented by the firing of a few cells regardless of position on the eye.

More recently Hubel and Wiesel have found that pattern information of various kinds is brought together in 'columns' arranged at right angles to the clearly visible layers of the striate cortex. These functional columns were discovered with subtle techniques as they are invisible to the eye. They seem to solve the problem of how the brain relates together not only three spatial dimensions but also colour, movement and other object-characteristics. A simple map would be quite inadequate, for there are insufficient spatial dimensions to represent more than the three spatial dimensions of external objects — since the brain is itself an object in normal three-dimensional space.

It seems from the electrophysiological data that perceptions are built from neural mechanisms responding to certain simple shapes,

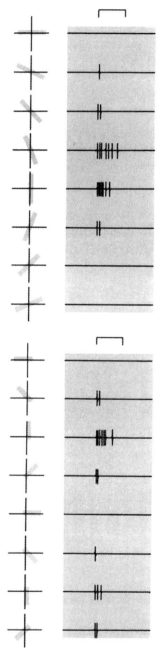

Fig. 12. This basic experiment by D. Hubel and T.N. Wiesel shows the activity of a single cell of the visual cortex of the brain as it is affected by a simple shape at the eye — a line at various orientations. The animal, a cat, views the (grey) line at various orientations. The selected cell whose activity is being recorded (shown by the spikes of electrical activity) fires at a certain orientation of the line at the eyes. Other cells respond to other orientations; others to movement; others to changing illumination. The features to which various cells respond form the 'words' of the brain's perceptual 'language', which is organised by what may be called the 'grammar of perception'. Electrophysiological experiments are beginning to tie up with psychological experiments. We begin to see how the pictures in the eyes are represented, or described, by brain activity.

Recently, Hubel and Wiesel have shown that many sensed characteristics are brought together in functional 'columns' in the brain, arranged at right angles to the layers of the *striate* area.

movement and colour. These are combined in the newly discovered cortical 'columns'. This is — logically — something like letters being combined to form words: the selected features are evidently basic units of the perceptual 'language' of the brain. What is not at all clear is how — to continue the language analogy — the neural 'words' are combined to form perceptual 'sentences'. It is not known at the neurological level how the outputs from the 'columns' are combined to give object-perception. We may guess that there is an intimate connection with memory stores, but at the present time how memory is stored is not known. It is not even known whether single cells store units of memory, or whether memories are stored as patterns involving very many cells, possibly by a process similar to the storing of optical patterns by holography and unlike the usual point-to-point representation as in normal photography. The answers to these questions remain for the future, but meanwhile it is worth preparing the ground by considering the phenomena of perception. Neurophysiological explanation is not everything. The activity of the nervous system cannot be interpreted without knowing what functions are served. Many useful explanations are in terms of function rather than in terms of underlying structure and activity within this structure. For example, the computer engineer does not have to know much physics to understand computer circuits; and the mathematician does not have to understand much electronics to understand their logic and use them. (To say, 'I understand why she went off with Bill', may be perfectly meaningful without any knowledge of what — in physical terms — went on in her brain.)

At this point we might be tempted into thinking that perception is simply a matter of combining activity from various pattern-detecting systems, to build up neural 'descriptions' of surrounding objects. But perception cannot be anything like as simple, if only because of a basic problem confronting the perceptual brain — the ambiguity of sensory data. The same data can always 'mean' any of several alternative objects. But we experience but one, and generally correctly. Clearly there is more to it than the putting together of neurally represented patterns to build perceptions, for decisions are required. We should look at the ambiguity of objects to see this more clearly. Establishing that a given region of pattern represents an object and not background is only a first step in the perceptual process. We are left with the vital decision: *What (kind of) object is this?*

The problem is acute because any two-dimensional image could represent an *infinity of possible three-dimensional shapes*. Often there are extra sources of information available; for example stereoscopic vision, or changing parallax as the head moves, but the fact remains that we can nearly always arrive at a reasonably reliable solution to the problem: 'What object is this?' Even though the number of possibilities is infinite.

Some shapes are seen, at different times, in more ways than one. Just as the object-space reversible figures spontaneously change, so some shapes though continuously identified as object yet spontaneously change as to *what* object it is, or what position it is being viewed from. Here we must discuss the work of the psychologist who has devised the most striking demonstrations based on the essential ambiguity of objects — Adelbert Ames.

Ames started out as a painter, but ended by devising many of the best known 'visual demonstrations'. It has not, however, always been made clear what they demonstrate. Unfortunately Ames himself wrote very little: he was a visual man.

Ames made several models (sometimes full scale) designed to give the same retinal image as familiar objects, though the models were in fact of very different shapes. The models gave the same image as the familiar object only from one critical view point, and for a single eye. The best-known demonstration is the 'Ames room'. This gives the same image to an eye placed at the critical position as a normal rectangular room — but in fact it is very far

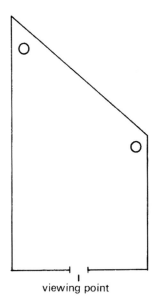

viewing point

Fig. 13.   The Ames room — how it is made. It is a highly non-rectangular shape, but so designed that it gives the same retinal image (or photograph) from a critical view point as a normal rectangular room. (A picture, indeed, is an extreme case of the same thing, the picture being flat though it may give the same image to the eye as normal three-dimensional objects.) The Ames room is not flat, but is a queer-shaped room such that with increasing distance there is a corresponding increase in size, to give the image of a normal rectangular room. If made correctly, it *must* look like a normal room, for it gives the same image to the eye. What is interesting is what happens when we put objects, especially people, inside it. Do they — or the room—look odd?

from rectangular. The further wall recedes to one side, so that one of the far corners is much further from the eye than the other corner; but both corners subtend the same angle at the eye placed in the critical position; for as the further wall recedes it gets correspondingly larger. The Ames room is simply one of an infinite set of three-dimensional shapes giving the same image to the (critically placed) eye that it would receive from a normal rectangular room.

As a matter of fact, Ames was not the first to consider such a situation. Helmholtz suggested such a room fifty years before when he wrote:

Looking at the (normal) room with one eye shut, we think we see it just as distinctly and definitely as with both eyes. And yet we should get exactly the same view in case every point in the room were shifted arbitrarily to a different distance from the eye, provided they all remained on the same lines of sight.

Ames was, however, probably the first actually to *make* such a 'distorted' room, and he was the first to consider what would happen if one placed objects of familiar size inside — actually at different but apparently the same distance from the eye. What happens is shown in figure 14. This is a view from the critical position of an Ames room, with two people inside. One looks smaller than the other though in fact they are the same actual height. The apparently smaller person is about twice as far from the camera as the other, and so the image at the camera (or at an eye in this position) is half the size. So this is *not* a case of a visual distortion illusion (cf. page 74) [not included here]. It is that we trade size and distance wrongly in this situation.

Clearly the room without the people or other objects in it *must* look like a normal rectangular room — for it gives the same image at the eye. Perhaps, indeed, Helmholtz decided that to make a 'distorted' room would merely demonstrate the obvious. But adding the objects inside changes the situation; for now the eye is presented with a betting problem: 'Is the *room* an odd shape, or are the *people* odd sizes?' It is an experimental result, not to be anticipated, that observers continue seeing the room as normal (which it is not) and the people as different heights (which they are not). The odds have been rigged, and the brain makes the wrong bet. It loses; so we are fooled. The odds are easily changed: it has been reported that a newly married

Fig. 14.   The Ames room — how it appears. The right-hand person looks much taller than the other. They are in fact the same height, but present different sizes to the eye. The further wall recedes, towards the left (as shown in figure 13). The left-hand person is twice as far from the camera as the other. Perspective is 'used backwards' to give the same retinal image as a normal rectangular room. So, apart from the figures, it *must* look the same as a rectangular room. (The question is, however: why does *any* room or other object look any *particular* shape?) This is an acute problem, because any projected image could be given by an infinity of differently shaped objects. The room is assumed (wrongly) to be rectangular, in which case the people must be very different in size. This shows how we base perception on bets. This is not a 'distortion illusion'.

wife will not see her husband shrink as he walks across the room, but instead will see the room more or less its true peculiar shape. No doubt this effect might be used for calibrating wives.

If the room is explored with a long stick, it will gradually come to look its true queer shape. Such active exploration with a stick will correct the visual perception — though intellectual knowledge of the true shape of the room will not. It is an effort to get perceptual and intellectual knowledge to coincide. If the eighteenth-century empiricists had known this, philosophy might have taken a very different course. No doubt there are also implications for political theory and judgement.

Is this all there is to the matter? Not quite. The Ames room is an interesting and dramatic demonstration — but is it an experiment from which conclusions can be drawn? If so, it is odd, for where is the control situation? Has the perceptual effect of the queer-shaped room really been demonstrated with no control?

The question to ask, surely, is: what happens in the Ames room *without the room?* In other words — if we take two people with a blank background, and place one at twice the distance of the other with no visual information of their relative distances — does the more distant one look the same distance but half the size? If not, what effect can we attribute to the Ames room?

Figure 15 shows two people, one placed at twice the distance of the other, so that their images differ in size just as in the Ames room picture. Also, a low camera angle has been used to prevent perspective information of their relative distances. This photograph is not a composite, and has not been touched up in any way. Most people looking at it say that the actually nearer person looks a little nearer, but also a lot larger. In other words, the size difference is *not* attributed purely to distance when the Ames room has been removed. So perhaps rather too much weight has been placed on what we may call the Ames addition to the Helmholtz situation.

Another demonstration is the Ames chair. This consists of several rods supported on thin wires, all converging to one point. The rods are viewed from this point of convergence, the wires following perspective lines from the eye to the rods. The rods are so arranged that, from this point of view, they form the image of a chair. They are then seen as a (model) chair. From any other point of view, of course, they are seen as a collection of disjointed rods. This is interesting, but like the room with nothing in it, it *has* to

Fig. 15.   The Ames-room-without-the-room! The people are photographed just as in the Ames room but without the room. One is twice as distant as the other. Is this how they appear? Most observers report that the girl photographed as smaller (actually most distant) appears shrunk, though also somewhat more distant. If people look shrunk when their image is smaller without the Ames room, we should re-evaluate the significance of the Ames room experiment.

work. If the model is made well enough, it *must* look the same as objects of different shape but giving the same image at the eye.

What is really remarkable is not so much that the Ames demonstrations work (in the absence of conflicting evidence, and the conflicting cases *are* very interesting), but that the perceptual system never does settle for one interpretation of retinal images from normal objects. Perhaps this is what the demonstrations bring out. It is surely remarkable that out of the infinity of possibilities the perceptual brain generally hits on just about the best one.

We are forced, at this early stage, to suppose that perception involves betting on the most probable interpretation of sensory data, in terms of the world of objects. Perception involves a kind of inference from sensory data to object-reality. Further, behaviour is not controlled directly by the data, but by the solutions to the

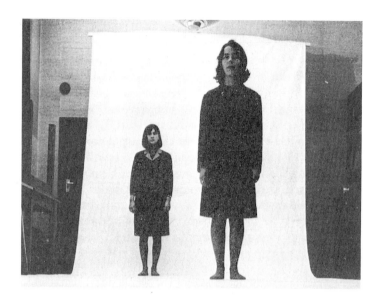

Fig. 16.   How the Ames-room-without-the-room picture was taken. The girls are actually at different distances from the camera, with a white background and a floor-level camera-view to reduce texture and perspective information of their distances. Before we know what happens when information is removed, we cannot assess the effects of misleading or other information on perception. It is vital to know what assumptions we make and how these are modified by retinal information in various circumstances.

perceptual inferences from the data. This is clear from common experience: if I put a book on a table I do not prod the table first to check that it is solid. I act according to the *inferred* physical object — table — not according to the brown patch in my eye. So perception involves a kind of problem-solving — a kind of intelligence.

Helmholtz spoke of perception in terms of 'unconscious inferences'. For reasons not altogether clear to the present writer, this has never been very popular among psychologists. Helmholtz was particularly concerned with the fact that though sensory activity starts at surfaces of the body, including the retina, we experience 'things out there'. Also, illusions of 'out-thereness' can be extraordinarily powerful. For example, a dark room lit by a bright electronic flash

Figs. 17 and 18.   The Ames chair: a collection of rods suspended in converging wires. From the point of convergence they give the same image as a chair — and so are seen as a chair — though from other positions they appear as a collection of disjointed rods.

will give an intense after-image in which every detail of the room is visible. Now although we know it is but a 'photograph' from the flash, on the eye, and though we turn round or walk out of the door with our after-image — yet we cannot see it as anything but a real room while the after-image remains. Helmholtz supposed, though he did not put it quite like this, that the brain continually carries out 'unconscious inference' of the form:

(Nearly) all retinal activity received is due to external objects.
This is retinal activity.
Therefore this is due to an external object.

At this point we must be clear that there is no 'little man inside' doing the arguing, for this leads to intolerable philosophical difficulties. Helmholtz certainly did not think this, but his phrase 'unconscious inferences', and his description of perceptions as 'unconscious conclusions' did perhaps suggest, at the time, to people unfamiliar with computers, some such unacceptable idea. But our familiarity with computers should remove temptation towards confusion of this kind. For we no longer think of inference as a uniquely *human* activity involving consciousness.

Helmholtz described his 'unconscious inferences' in perception in the following words, which I quote at length:

The psychic activities that lead us to infer that there in front of us at a certain place there is a certain object of a certain character, are generally not conscious activities, but unconscious ones. In their result they are equivalent to a conclusion, *to the extent that the observed action on our senses enables us to form an idea as to the possible cause of this action; although as a matter of fact, it is invariably simply the nervous stimulations that are perceived, that is, the actions, but never the external objects themselves. But what seems to differentiate them from a conclusion, in the ordinary sense of the word, is that a conclusion is an act of conscious thought. An astronomer, for example, comes to real conclusions of this sort, when he computes the positions of the stars in space, their distances etc., from the perspective images he has had of them at various times and as they are seen from different parts of the orbit of the earth. His conclusions are based on conscious knowledge of the laws of optics. In the ordinary acts of vision this knowledge of optics is lacking. Still it may be permissible to speak of the psychic acts of ordinary perception as* unconscious conclusions, *thereby making a distinction of some sort between them and the so-called conscious conclusions. And while it is true that there has been, and probably always will be, a measure of doubt as to the similarity of the psychic activity in the two cases, there can be no doubt as to the similarity between the results of such unconscious conclusions and those of conscious conclusions.*

This book may almost be regarded as an extension of this passage from Helmholtz. It is clearly first of all vital for an animal to distinguish, from the patterns in its eyes, between what are objects in the field of view and what is the space between objects. It is then necessary for it to identify the objects, from their characteristic patterns. But, as we have said, objects are far more than patterns at the senses. And it is these other, non-sensed, characteristics which are important to the owners of eyes. Are objects inferred from patterns? In this book I propose to consider the inner 'logic' of perception. The main argument is that perception is a kind of problem-solving. Pictures are regarded as a remarkable invention, requiring special perceptual skills for seeing them, leading to abstract symbols and ultimately to written language. By considering the perception of objects represented in pictures and symbols (including the pictograms of early languages) I hope to show that our most abstract thinking may be a direct development of the first attempts to interpret the patterns in primitive eyes in terms of external objects.

Gallery of Monsters

Ian Stewart*

The eternal battle between order and disorder runs like a deep ocean current through the human perception of the universe, a common feature of many creation myths from many cultures. In the Old Testament, 'the Earth was without form, and void, and darkness was upon the face of the deep'. In the Enuma Elish, an early Babylonian epic, the universe arises from the chaos that ensues when an unruly family of gods is destroyed by its own father. Order is equated with good and disorder with evil. Order and chaos have always been seen as two opposites, twin poles about which we pivot our understanding of the world.

Not any longer. Today, at the frontiers of humanity's exploration of the magical maze, we are discovering that order and chaos are not opposites, but soulmates — two sides of the same coin, two edges of the same sword. A popular version of this new viewpoint has leapfrogged the tiresome process of objective scientific testing, establishing itself as 'chaos theory' — a world of psychedelic posters and postcards, anarchistic philosophies, maverick guru scientists, and devastating hurricanes caused by one flap of a butterfly's gossamer wing.

* The original version appeared in: Ian Stewart, Passage 8, *The Magical Maze: Seeing the World through Mathematical Eyes*, London, 1997, pp. 215–247.

The media have attributed almost mystical powers to chaos theory. When they're not rubbishing it, that is.

There is a mathematician in *Jurassic Park,* called Ian Malcolm, and he is a chaos theorist. He knows that Jurassic Park's complex systems are doomed, right from the start, but nobody listens until it's too late. This is chaos theory as Cassandra. In fact, the movie doesn't actually tell us much about chaos, and most of what it does say misses the point, which is about what you'd expect from Hollywood. The book version is better, and it avoids the trite 'Man was not intended to meddle' ending. Not that the book tells you much about chaos theory either.

So I guess it's up to me☛.

Chaos theory has two main ingredients: geometrical shapes known as 'fractals', and irregular behaviour called 'chaos'. The two concepts evolved separately, but have since become inseparably intertwined. Chaos theory sheds new light on the predictability of nature — and also muddies the waters. Cassandra it is not, but it sometimes sounds just like her. Chaos theory provides a new angle on an old discovery: 'laws of nature'. The revolution of scientific thought that culminated in Isaac Newton led to a vision of the universe as some gigantic mechanism, functioning 'like clockwork', slavishly obeying simple, fixed mathematical principles. Philosophers call this idea 'determinism'. In particular, it led Pierre Simon de Laplace — one of the great eighteenth-century mathematicians — to an astonishing vision of a vast intellect, capable of predicting the behaviour of the entire universe from a single formula. In a fully deterministic universe, Laplace pointed out,

> an intellect which at any given moment knew all the forces that animate nature and the mutual positions of the beings that comprise it, if this

---

☛For a more extensive discussion of nearly everything in this final passage of the magical maze, and a lot more, see Ian Stewart, *Does God Play Dice?* (2nd edn), Penguin, Harmondsworth, 1997.

intellect were vast enough to submit its data to analysis, it could condense into a single formula the movement of the greatest bodies of the universe and that of the lightest atom: for such an intellect nothing could be uncertain, and the future, just like the past, would be present before its eyes.

In Douglas Adams' *The Hitch Hiker's Guide to the Galaxy*, Laplace's vision was parodied in the ultimate supercomputer Deep Thought, with its answer 'forty-two' to the great question of Life, the Universe, and Everything. In Laplace's disembodied super-intellect we find the paradigm of the clockwork universe, never deviating from its initial course once its cogwheels have been set in motion. And, for all its faults, it has been spectacularly successful in helping humanity to come to terms with the world around it.

But now, mathematicians and physicists have discovered something rather curious. Order can breed its own peculiar kind of disorder. Deterministic causes — equations that do not contain any random terms, which in principle describe the evolution of some system uniquely for all time — can have random effects.

This is a remarkable discovery, and it is changing the face of science. It is known as deterministic chaos, or just plain chaos. Chaos lies at the frontiers of today's mathematics, one of several startling new paradoxes about the way the world can change. Others include the self-organisation of evolutionary systems, their self-complication, the spontaneous emergence of 'computational' entities, and the emergence of large-scale order from small-scale disorder. Nature is far more complicated, far more interesting — and far more clever — than we think. Its patterns are not the direct consequences of simple laws, but emerge indirectly from an all-embracing sea of chaos and complexity.

Take your own heart. Traditional science treats it as a pump, beating 'like clockwork', whose cycles can be dissected into simple waves of standard shapes. Real hearts are far more puzzling. Your heartbeat is triggered by signals from your brain, but the rhythmic contractions that keep you alive are the result of a democratic vote by millions of muscle fibres, all agreeing to contract in synchrony. Like the flashing fireflies.

A simple, clockwork process?

Even when your body is at rest, your heartbeat varies by tiny but measurable amounts. This variability is not the result of random outside influences: it is caused by chaotic internal dynamics. There are good reasons for the chaos: your heart wouldn't work without it. Chaos distributes wear and tear more evenly; it makes your heart able to respond more rapidly to changes in your surroundings. A clockwork heart would only work in a clockwork person.

THE CHAOS GAME

A parallel paradox is that genuinely random processes can lead to a surprising degree of order. In his book *Fractals Everywhere,* the British-born mathematician and entrepreneur Michael Barnsley introduces what he — rather confusingly — calls the Chaos Game.

Really, it should be called the Fractal Game — but who am I to contradict its inventor? After all, look what happened to the people who disagreed with the mathematician in *Jurassic Park.*

The Chaos Game is both simple and surprising. In a daring act of imagination, Barnsley has parlayed its simplicity, and its surprise, into a multi-million dollar industry. Behind the Chaos Game lie ideas that have transformed the way we store and transmit visual images. Related ideas have given mathematicians a completely new way to model nature, one whose implications are only just beginning to be recognised.

To play the Chaos Game, you need a piece of paper, a ruler, a pencil, and a 'three-sided coin' for which 'heads', 'tails', and 'edge' have the same probability, 1/3. For example, you could use a die, and let 1 or 2 = heads, 3 or 4 = tails, and 5 or 6 = edge. Mark three points on the sheet of paper, say at the vertices of an equilateral triangle. Label the three corners of the triangle 'heads', 'tails', and 'edge'. Use your pencil to mark a random point on the paper. Toss the 'coin', and move the point half-way towards the corresponding vertex, getting a new point. Draw that, too. Repeat this procedure, starting from the new point, and always generating the next point from the previous one by tossing the 'coin' and moving half-way towards the appropriate vertex (Figure 73).

What do you see?

You might expect the result to be some uniform cloud of points in the plane. Not so. Figure 74 shows what happens with a thousand trials, drawn using a computer.

This strange shape is the *Sierpiński gasket,* and we've already encountered something very like it in the chapter 'Panthers Don't Like Porridge', in connection with the Tower of Hanoi. The gasket was invented by the Polish mathematician Waclaw Sierpiński: for him it was an example of a curve that crosses itself at every point. The gasket is constructed from an equilateral triangle. Divide it into four equal quarters, and throw away the middle one. This leaves three triangles half the size. Divide each into four equal quarters, and throw away the middle one. This leaves nine triangles one-quarter the original size. Repeat for ever: the gasket is what remains

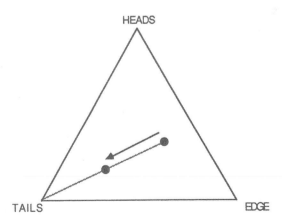

Figure 73   How to play the Chaos Game.

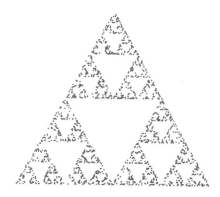

Figure 74   A thousand trials of the Chaos Game. Familiar?

Figure 75   Stage five of a Sierpiński gasket.

when everything else has been thrown away. Figure 75 shows what it looks like after the first five stages of its construction.

The Sierpiński gasket seems a very odd shape to be generated by a random procedure. In fact, it seems a very odd shape — period. When such shapes first appeared on the mathematical scene, about a century ago, they were derided as a 'gallery of monsters'. At that time their main purpose was to shine a light into the murkier corners of the magical maze by showing how nasty mathematics could get. Several leading mathematicians saw little point in doing that, and said so — often with some sarcasm.

Today, however, we recognise that shapes such as Sierpiński's are entirely natural — and useful. His gasket is one of the more regular representatives of a class of mathematical shapes called *fractals*. Fractals were invented, named, and promoted by Benoît Mandelbrot, and they are a new way of modelling the irregularities of nature.

BONSAI MOUNTAINS

The shapes studied in classical geometry are things like triangles, squares, lines, circles, ellipses, spheres, and so on. They are all very simple shapes, and they share a common feature: they have no interesting structure on sufficiently small scales.

Consider a circle, for example. Imagine looking at it through an enormously powerful microscope. In the field of vision of the microscope you see a tiny portion of the circle, magnified to an immense size. When the magnification is low, you see a curved circular arc. As the power is turned up, you still see a circular arc, but at high scales of magnification it no longer looks curved. It is curved, but the curvature is so slight that the eye no longer notices it. At high magnification, a circle looks almost exactly like a straight line, with no interesting features whatsoever.

The same goes for a sphere: at high magnification, a sphere looks almost exactly like a plane, again with no interesting features whatsoever. This is why primitive cultures thought the Earth was flat. It *looks* flat when all you can see is a few square kilometres of its vast, gently rounded surface.

When you magnify a Sierpiński gasket, however, it does not become featureless. Inside the gasket are triangular holes, as tiny as you wish. However much you magnify the gasket, the holes will still be there. Holes so tiny that your eye cannot see them will grow until they almost fill the field of vision. In fact, no matter how much you magnify a Sierpiński gasket, what you see will always look much like a Sierpiński gasket. Unlike the sphere

Figure 76   A Sierpiński gasket is made from three perfect copies, each half as big.

and circle, the Sierpiński gasket has fine detail at any scale of magnification.

In fact, the Sierpiński gasket is built from three identical copies, each one half the size (Figure 76). Each of those is again made from three identical copies, each one half the size — so the Sierpiński gasket is built from nine identical copies, each one-quarter the size, too. By repeating this argument we see that it can be made from 27 identical copies, each 1/8 the size; 81 identical copies, each 1/16 the size; 243 identical copies, each 1/32 the size . . . and so on indefinitely. This proves that we can never run out of detail, because however small the piece you look at, it will contain a tiny, perfect copy of the entire gasket. We say that the gasket is *self-similar:* tiny bits of it look much like the entire object.

Many shapes in nature have the same kind of delicate, hidden, self-similar detail. Unlike a true mathematical fractal, whose detail goes on for ever, real shapes do eventually fuzz out when we get down to atomic scales — but, even so, they are much better modelled by a fractal than they are by a sphere or a circle.

Mountain ranges have structure on very small scales. A mountain range is a collection of jagged peaks. Each peak is itself a collection of jagged sub-peaks, and so on. A lump of rock broken off a mountain looks very much like a miniature mountain: this is why Terry Pratchett's book *Small Gods* can feature a character whose hobby is 'bonsai mountains'. You'll recall that the Japanese art of bonsai produces miniature trees by growing small shrubs and training them into the shape of a mature tree. The process takes years, and uses all sorts of special equipment and techniques. Bonsai is possible because a small part of a tree has a very similar structure to the whole tree. Pratchett's six-thousand-year-old 'history monk' Lu-Tze can culture bonsai mountains — using equally specialised equipment, such as tiny mirrors to simulate the glare of the sun and a watering-can to simulate the erosion caused by thunderstorms — because a jagged lump of rock retains all of the complexity of a huge mountain range. After all that practice, he does it extremely well. As the novice Brutha asks, 'That can't really be snow on the top, can—' at which point he is interrupted by the Great God Om, who is a small tortoise.

Coastlines have the same property as bonsai mountains. If you look at a map of a county, country, or continent that possesses a coast, you'll find that the coastline always looks much the same — not in detail but in 'texture' — whatever the scale of the map. Maps drawn to a larger scale show more detail, but the general nature of that detail is always the same: bays, promontories, random-looking wiggles. Coastlines are 'statistically self-similar': little bits of them look like big bits of some *other* possible coastline.

Many plants behave in the same manner, too. I've already mentioned trees. Ferns are beautifully fractal: a fern leaf is made from a series of fronds, sticking out to the left and right from a central spine. Each frond is made from a series of sub-fronds, each sub-frond from a series of sub-sub-fronds, and so on — for four of five steps, typically. Even more striking is broccoli romanesco (Figure 77), a cauliflower-like plant. Its head consists of a spiral swirl of florets. Each floret is a

Figure 77   The fractal self-similarity of broccoli romanesco.

spiral swirl of sub-florets, each sub-floret is a spiral swirl of sub-sub-florets . . . and so on.

Bonsai coastlines, bonsai ferns, bonsai broccoli. All of them are far better modelled by fractals than by sphere, cones, cubes, pyramids, or the other paraphenalia [*sic*] of Euclid's geometry.

FRACTAL DIMENSION

In order to be able to do science using fractal models, we need a way to characterise a fractal quantitatively. We need a number that can be measured in an experiment, and captures some useful flavour of the fractal's geometry. The most important such quantity is known as the fractal dimension.

We'll sneak up on fractal dimension by thinking about more traditional shapes — lines, squares, cubes. In ordinary mathematical parlance, a line is 1-dimensional, a square is 2-dimensional, and a cube is 3-dimensional. One way to see this is to notice that a line pokes out along one direction (say east), a square pokes out along two independent directions (east and north), and a cube pokes out along three directions (east, north, up). But there's different way to work out the dimension, which uses scaling.

Suppose you want to make a line twice as big. You can do so by joining together two copies (Figure 78(a)). If you want to make the line three times as long, you can do so by joining together three copies (Figure 78(b)). And so on. To make the line n times as long, you join together n copies.

Next, suppose you want to make a square twice as big. Then you can do so by joining together *four* copies (Figure 78(c)). If you want to make the square three times as big, you can do so by joining

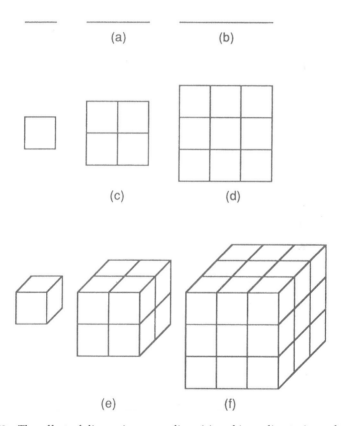

Figure 78   The effect of dimension on scaling: (a) making a line twice as big, or (b) three times as big; (c) making a square twice as big, or (d) three times as big; (e) making a cube twice as big, or (f) three times as big.

together nine copies (Figure 78(d)). To make the square n times as big, you join together n^2 copies.

Continuing up the dimensions, suppose you want to make a cube twice as big. Then you can do so by joining together *eight* copies (Figure 78(e)). If you want to make the cube three times as big, you can do so by joining together twenty-seven copies (Figure 78(f)). To make the cube n times as big, you join together n^3 copies.

A square is the 2-dimensional analogue of a cube; a line segment is the 1-dimensional analogue. Suppose we choose a dimension d. Then we can summarise our findings on the scaling of d-dimensional 'cubes' thus: to make a d-dimensional 'cube' n times bigger, you need n^d copies.

We can 'solve' this statement for d, if we're prepared to use logarithms☞. Let

$$k = n^d$$

be the number of copies. Taking logarithms, we get

$$\log k = d \log n,$$

so

$$d = (\log k)/(\log n).$$

The dimension of a 'cube' is the logarithm of the number of copies, divided by the logarithm of the size.

Suppose we apply all this to the Sierpiński gasket, without worrying too much (yet) about what the calculation signifies. We saw that the Sierpiński gasket can be doubled in size by taking

---

☞ If $x = e^y$, where $e = 2.718 \ldots$ is the base of natural logarithms, then $y = \log x$. The main property of logarithms that we use here is

$$\log (ab) = \log a + \log b,$$

together with its consequence

$$\log(a^c) = c \log a.$$

three identical copies. So $n = 2$, $k = 3$. Plugging those numbers into the formula for d, we find that the dimension of the Sierpiński gasket is

$$d = (\log 3)/(\log 2),$$

which is about 1.584962.

The dimension of the Sierpiński gasket is not a whole number!

Well, that certainly explains why it looks so strange. 'Gallery of monsters', indeed.

You could choose to dismiss the calculation as meaningless: how can a Sierpiński gasket 'poke out along 1.58 directions'? But the imaginative mathematician follows his or her nose when they blunder into a new corner of the magical maze. And the trained mathematical nose, in this instance, smells something interesting. For what we are witnessing is this: the traditional concept of dimension can be extended to fractals, provided we focus on scaling properties, and *not* on 'number of directions'. The former generalises to fractals, the latter does not.

We say that the Sierpiński gasket has *fractal dimension* 1.58. The interpretation is that it 'scales' in a manner that lies somewhere between what a 1-dimensional object would do and what a 2-dimensional object would do. It takes two copies of a line to double its size, and four copies of a square to double its size. In between is the Sierpiński gasket, for which only three copies are needed. So it's entirely reasonable that its 'dimension' should lie between 1 (the dimension of the line) and 2 (the dimension of the square).

Figure 79 shows another fractal, the Sierpiński carpet. Here a square is repeatedly divided into nine sub-squares, each one-third the size, and the central sub-square is removed. What is the fractal dimension of the Sierpiński carpet? Well, it takes $k = 8$ copies to make the size increase by a factor of $n = 3$. So the fractal dimension is

$$d = (\log 8)/(\log 3) = 1.892789.$$

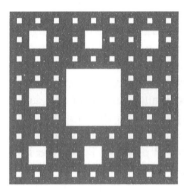

Figure 79   The Sierpiński carpet.

Notice that this also lies between 1 and 2, but now it is a bit bigger than the dimension of the Sierpiński gasket. This is again reasonable, because the Sierpiński carpet is distinctly less 'holey' than the gasket. The fractal dimension captures 'how well the shape fills space', or perhaps 'how irregular it is'.

FAMILY TREES

Fractals can be used for all sorts of purposes. Scientists use them to model the growth of loosely knit clusters, such as soot; here computer models of the growth process generate fractals (Figure 80) whose dimension is very close to that measure in real soot. The scientist concludes that the model growth process bears some degree of resemblance to the real one, and so learns more about the formation of soot. Metallurgists use fractals to understand crystal growth in solidifying metals. Meteorologists use fractals to understand clouds. Very recently, a team of medical researchers discovered that they could explain some puzzling numerical features of blood flow if they assumed a fractal model for the body's veinous system.

Here, I want to show you a little of how fractals model plants. The idea is to set up simple rules that generate the branching pattern of the plant, and then represent the branches geometrically.

Figure 80 Fractal model of soot.

The result is a self-similar branching structure, a fractal plant. (It is self-similar in that any branch broken off obeys the same rules, and so is a smaller version of the same 'species'.) The rules are called *Lindenmayer systems* after their inventor, Aristid Lindenmayer, or L-systems for short.

One of the simplest L-systems arises in connection with Fibonacci's rabbit problem. Remember how that goes. We start with a pair of immature rabbits. Each breeding season, the immature pairs mature for a season, while each mature one breeds a new immature pair. Earlier we asked how fast the population grows, but now we are going to look at the shape of the rabbits' family tree. 'Tree' starts out by being a metaphor here — and ends up a lot closer to reality. Fibonacci numbers don't just describe the spirals in sunflower heads, they are the visible tip of a marvellous mathematical theory of branching structures. The same scheme illuminates not just the numerology of plants, but their entire form — the way they, too, branch. We can now — in computer simulation — *grow* realistic grasses, flowers, bushes, and trees from mathematical rules. And there is a strong suspicion that those rules lie at the heart of how plants themselves grow.

We can summarise the growth pattern of the rabbits by using a 'grammar' in which the letters I and M stand for immature and mature pairs, respectively. Then the growth rules lead to a system of transformations, which take us from one season to the next:

$$I \rightarrow M \quad \text{(immature pairs mature after one season)}$$
$$M \rightarrow MI \quad \text{(mature pairs stay alive and breed a new immature pair)}.$$

We start with one immature pair (I) and repeatedly apply the two branching rules to get the sequence of symbol strings

$$I \rightarrow M \rightarrow MI \rightarrow MIM \rightarrow MIMMI \rightarrow MIMMIMIM \rightarrow \ldots$$

At each step we apply the branching rules to every symbol in the string, and this leads to the next string. For example, to go from MIM to MIMMI we do this:

$$
\begin{array}{ccc}
M & I & M \\
\downarrow & \downarrow & \downarrow \\
MI & M & MI.
\end{array}
$$

Incidentally, we can solve the original Fibonacci problem 'how big is the rabbit population?' using this 'grammar', just by counting the symbols. For example, the final generation listed contains five M's and three I's, a total of eight pairs of rabbits. These are three consecutive Fibonacci numbers, and the pattern continues indefinitely.

Instead of counting the symbols, we can *interpret* them as branches in a tree diagram (Figure 81). Now we are modelling a plant which has two cell-types: immature ones, which mature for a season and then branch, and mature ones, which produce an immature side-branch while continuing to grow themselves. No longer a family tree of rabbits, this is just a tree — or some tree-like

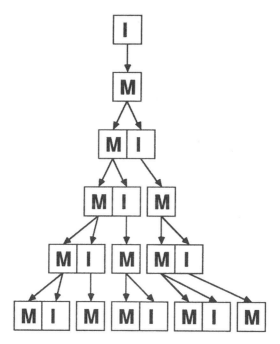

Figure 81   The diagram determined by Fibonacci's rabbits.

plant. If, say, the immature cells are round blobs and the mature ones are long and thin, we get a plant something like Figure 82.

L-systems help to explain the patterns in which plants branch. If you look closely at plants and shrubs, you will often find that their branches don't divide at random but have regularities. Perhaps branching will lead to one long stem and one short one, each of which branch in turn according to the same pattern, for instance. Figure 83 shows some plant-like shapes generated by various L-systems.

So, by a curious quirk of fortune, it turns out that branching patterns in plants, as well as numbers of petals, depend on mathematics that arose from Fibonacci's rabbit problem. But now it is the rabbit's family tree, not the size of their population, that matters.

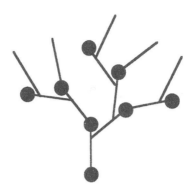

Figure 82   A plant designed using the rabbits' family tree.

MATHEMATICS CAN BE FERN

Barnsley's image processing methods also stemmed (pun intended) from a plant — the black spleenwort fern. This beautiful fractal (Figure 84) is made from four slightly distorted copies of itself. Musing on the how such remarkable complexity can be generated by simple fractal rules, Barnsley suddenly realised that the fern contained the germ of something potentially very important: image compression.

In today's world, we spend a lot of time and money to send images all over the globe. Television — terrestrial, satellite, cable — does little else, at 25 images per second. Many images are transmitted over the Internet. Businessmen fax pictures to their distant colleagues.

When you fax someone an image, the fax machine scans it, row by row, and turns the picture into a long series of binary signals — 0 for white, 1 for black, say. (The actual process is more sophisticated, but this description will do.) Then the signals are sent along a phone line as a series of bleeps. At the other end, the receiving fax decodes the bleeps and turns them back into black and white dots, which the human eye sees as a picture. TV works the same way, but with colour information included too.

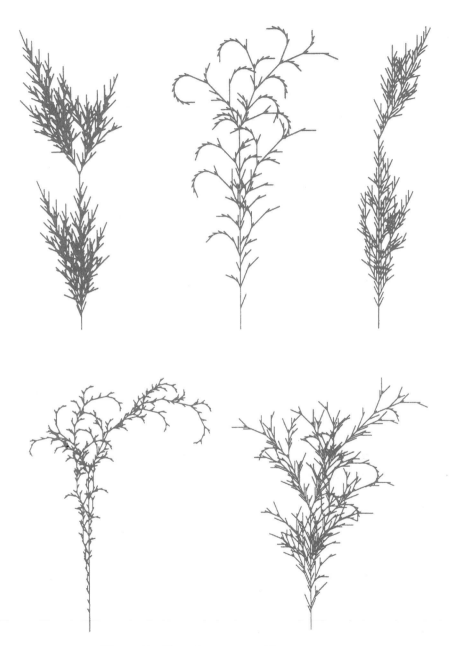

Figure 83 Plant shapes created by L-systems.

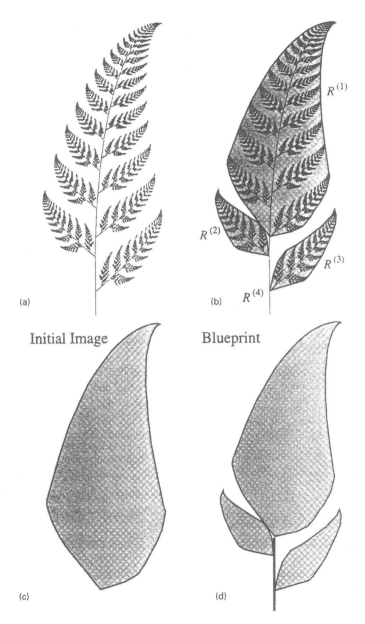

Initial Image   Blueprint

(a)   (b)   (c)   (d)

Figure 84 The black spleenwort fern. (a) The fern fractal. (b) The four transformations $R^{(1)}$, $R^{(2)}$, $R^{(3)}$, $R^{(4)}$, that generate it. $R^{(4)}$ squashes the whole fern flat on top of its stalk. (c) Initial course outline of fern. (d) The four transformations pictured as a geometric 'blueprint'.

One picture generates an awful lot of dots. At, say, 300 dots per inch, a picture the size of a postage stamp requires about 100,000 dots. A full-page image takes perhaps 10 million.

Suppose you want to send someone a full-page picture of the black spleenwort fern, by fax, using the conventional approach. Then you need to send 10 million binary digits. But — and this is the big insight — you can transmit the fractal *rules* for constructing the fern using only a few hundred binary digits — one-thousandth of 1% of the information. Of course, the person at the other end has to know how to turn those rules into a picture: this is where the Chaos Game comes in, but the details can wait for a few pages.

This reduction of information — the technical term is 'data compression' — also reduces transmission time, and hence cost. If you could compress video images by this amount, then every satellite TV channel could be replaced by 100,000 channels.

Of course, it's not *that* easy. TV involves sound as well as images. It takes time to turn an image into its fractal rules, and it takes more time to reverse that process at the other end. And, crucially, viewers want to look at things other than the black spleenwort fern. Soaps, sitcoms, blockbuster movies, Formula One, golf . . . Fortunately, the compression procedure is quite general: it works on any fractal. The world, as we've seen, is full of fractals. Most soaps and sitcoms have plants in the picture. Blockbuster movies happen in jungles, on mountains, or on the cratered surfaces of alien planets. Formula One has big cars in the foreground — and smaller ones, of much the same shape, in the background, trying desperately to catch up with the leader. Golf has trees all over the place, often between the ball and the hole. So Barnsley's idea was a promising start, and further research would surely improve it.

Barnsley tried to sell his marvellous idea to various major communication companies. No luck. It was too radical. So he set up his own company. And, after a lot of effort and a great deal of worry, he discovered that a variation on his idea works for *any image whatsoever.* The degree of compression is no longer by a factor of 100,000, mind you, but compression by a factor of 100 or so — reducing the picture information to 1% of its original size — is entirely feasible.

The method now works like this. A computer scans the image looking for parts of it that resemble other parts, but on a larger scale. If the computer can assemble a long enough list of such related pairs of parts of the image, then the entire image can be reconstructed from that list. The information needed to transmit the list is far less than that needed for the image itself. Figure 85 shows an image compressed using Barnsley's technique. The Encarta™ CD-ROM encyclopaedia contains 8,000 full colour images, compressed using Barnsley's fractal methods. Without image compression, it would have been more like 80 images.

How do we reconstruct an image from fractal rules?

Let's think about the fern. The rules boil down to a list of *transformations*, here four in number. Each transformation tells you how to take the plane, distort it, and map it on to itself. The fern is made up from four transformed copies of itself, using those four transformations.

It is, in fact, the *only* shape that is made up from four transformed copies of itself using those four transformations. In principle, then, knowing the transformations implies that you know the shape.

How, though, can you actually *find* the shape from the list of transformations?

Here's a really neat, clever method. Call the four transformations T_1, T_2, T_3, T_4, say. Imagine putting them into a bag. Choose some point P_0 in the plane, and draw a dot. Now begin pulling transformations out of the bag, at random.

Say you pull out T_3 first. Apply that transformation to P_0 to get a new point:

$$P_1 = T_3(P).$$

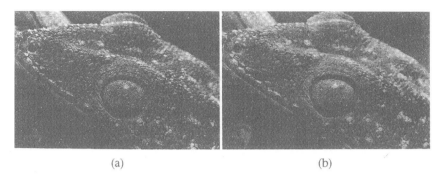

(a)   (b)

Figure 85   Image compression by Barnsley's method: (a) image of a gecko, uncompressed; (b) the same image compressed to require 1/156 as much memory.

Draw that point in the plane too. Put T_3 back in the bag, and again pull out a transformation at random — say, T_4. Now apply that transformation to P_1 to get a new point:

$$P_2 = T_4(P_1).$$

Draw that point in the plane.

Continue this process, pulling out a transformation at random, applying it to the last point drawn to get a new point, drawing that point on the plane, and putting the transformation back in the bag. You slowly build up a fuzzy cloud of points. After a while, the cloud starts to look suspiciously like the black spleenwort fern. The longer you continue the process, the closer the resemblance gets.

There are other ways to reconstruct the fractal from its rules, but this method is quick, simple, and surprising.

What of the Chaos Game? That is the same method applied to the Sierpiński gasket. The gasket is formed from three copies of itself. To get these copies, choose one of the three corners of the equilateral triangle and shrink the entire plane towards that point, so that distances all become halved. That's one copy: the other two come from the other two corners. The rules for the Chaos Game correspond to these three transformations. The choice 'heads', 'tails', or 'edge' is like picking a transformation out of the bag. And the process of drawing dots, and moving them half-way towards the randomly chosen corner of the triangle, is exactly the one outlined above. So in fact it is no surprise to find that the Chaos Game draws a Sierpiński gasket.

That's what it was designed to do.

PASCALS EVERYWHERE

Fractals turn up in the most unexpected places. For example, they can be found in Pascal's triangle, named after the seventeenth-century mathematician-cum-philosopher Blaise Pascal. He doesn't really deserve all the credit: the idea was already known a lot earlier. But he does deserve *some* credit, because he developed some important applications of his triangle, among them some basics of probability theory. Pascal's triangle looks like this:

$$
\begin{array}{ccccccccccccc}
 & & & & & & 1 & & & & & & \\
 & & & & & 1 & & 1 & & & & & \\
 & & & & 1 & & 2 & & 1 & & & & \\
 & & & 1 & & 3 & & 3 & & 1 & & & \\
 & & 1 & & 4 & & 6 & & 4 & & 1 & & \\
 & 1 & & 5 & & 10 & & 10 & & 5 & & 1 & \\
1 & & 6 & & 15 & & 20 & & 15 & & 6 & & 1 \\
\end{array}
$$

and so on. It arises in connection with the 'binomial theorem' in algebra: for example, if you work out

$$(x + y)^3 = x^3 + 3x^2y + 3xy^2 + y^3,$$

then the coefficients 1, 3, 3, 1 form the third row of Pascal's triangle. But that's by the by: what concerns us here is a beautiful pattern in Pascal's triangle, which lets us extend it as far as we wish. Each number in it (apart from the border of 1's) is equal to the sum of the two numbers one row above it, to the left and right. For example, the first 15 in row six is flanked, one row higher, by 5 and 10, like this:

$$5 \quad 10$$
$$15$$

and $5 + 10 = 15$. Using this rule, we can create Pascal's triangles with a thousand rows, or a million — given enough time and a fast computer.

Now, some of the numbers in Pascal's triangle are even, and some of them are odd. Obviously. But how can you tell which? The answer is elegant, and extremely surprising. To give you a hint of how strange it all is, here's a simpler question. If you drew a really big Pascal's triangle, say the first thousand rows, what proportion of the numbers in it would be even? About half? After all, up to any given limit half the numbers are even and the other half odd.

No. A lot more than half the numbers in the triangle are even. In fact, the larger the triangle, the closer the proportion comes to 100%. *Almost all* numbers in Pascal's triangle are even. Looking at a small triangle, this doesn't seem very likely, but I'll convince you that it must be true.

It helps to recast the problem in geometric terms. Draw Pascal's triangle as a grid of squares, like bricks in a triangular wall. Colour a square black if the corresponding number is odd, and white if it is even. We don't need to work out the exact numbers in Pascal's triangle in order to find out the colours. All we need is the symbolic rule that each number in the triangle is the sum of the two above it to the left and right:

$$a \qquad b$$
$$a + b$$

together with the information that

> 1 is odd,
>
> odd + odd = even + even = even,
>
> odd + even = even + odd + odd.

Then the rule can be thought of as telling us how to get the colour of one brick from those of the two above:

white   white
white

and

white   black
black

and

black   white
black

and

black   black
white.

So all we have to do is colour all the squares along the two sides of the triangle black, and then colour squares white if the two above them are the same colour, black if not. It doesn't take long to fill in quite a big Pascal's triangle.

There's another, more sophisticated, way to describe all this. A number is even if it is congruent to zero (mod 2), and odd if it is congruent to 1 (mod 2). Because all the usual rules of algebra apply in modular arithmetic, you can build the entire triangle just using arithmetic (mod 2), following exactly the same addition rule.

Be that as it may, what do we actually get? We get a dramatic and intricate pattern of black and white upside-down triangles

Figure 86   Pattern of odd and even numbers in Pascal's triangle.

(Figure 86), which looks just like the Sierpiński gasket. This is no coincidence: it can be proved that the Sierpiński gasket pattern persists no matter how big a Pascal's triangle you draw.

Odd and even numbers occur equally often (give or take the odd extra one) in any selected range of whole numbers. You'd be forgiven for assuming that the same is true of the numbers in Pascal's triangle: half even, half odd. However, the probability of getting an even number in Pascal's triangle is the proportion that is coloured white in Figure 75, and the probability of getting an odd number is the proportion coloured black. For larger and larger numbers of rows in Pascal's triangle, these two probabilities are approximated better and better by the corresponding proportions of a Sierpiński gasket.

So what proportion of a Sierpiński gasket is white?

Consider how the gasket is constructed. Start with a black triangle of total area 1. Paint an upside-down triangle, one quarter the size, white. This leaves three smaller black triangles, each of area 1/4, and the remaining black area has shrunk from 1 to 3/4. Now paint an upside-down white triangle on each of these: the black area shrinks to $3/4 \times 3/4$. Repeat indefinitely. More and more of the gasket gets painted white, and the black area becomes $3/4 \times 3/4 \times \ldots \times 3/4$, which tends to *zero* as the number of stages becomes very large.

In other words, the black part of a Sierpiński gasket has total area 0, the white part has area 1. This means that *almost all numbers in Pascal's triangle are even*. So we've learned something surprising about Pascal's triangle, by thinking fractally about the Sierpiński gasket.

CALCULATOR CHAOS

We saw that Barnsley's Chaos Game also generates a Sierpiński gasket, but for a rather different reason. It's a pity he used the name 'chaos', though, because what his game uses is randomness. Chaos *looks* random — but it's not generated by a random process.

It is, however, closely related to fractals. Chaos is highly irregular dynamic behaviour generated by non-random rules. Fractals are the geometry of chaos — its visual manifestation.

Chaos arises in 'dynamical systems' — systems in which things change over time according to a fixed mathematical rule. The rule is of the following general type: 'If the state of the system now is *this,* then its state one instant into the future can be determined from *this* by calculating *that.*' At first sight, any such system is totally predictable. Given its state right now, we apply the rule to find what it will do one instant into the future. Then we apply the rule a second time to see what it will do one instant after that — two instants into the future. Then we apply the rule a third time to see what it will do one instant after that — three instants into the future.

This is what led Laplace to his rather ambitious vision of universal determinacy. However, dynamical systems are not as predictable as they first appear, and this is where chaos comes to the fore.

Chaos isn't anything exotic. It happens all the time — the only novelty is that nowadays we *notice* it. You can observe chaos on a pocket calculator.

Many calculators have an x^2 (square) button. (If not, $\times$ followed by = has the same effect.) Pick a number between 0 and 1, such as 0.321, and hit the x^2 button. Do it again, over and over, and watch the numbers. What happens?

They shrink. By the ninth time I hit the button on my calculator, I get zero, and since $0^2 = 0$ it's no surprise that after that nothing very interesting happens.

This procedure is known as iteration: doing the same thing over and over again. A dynamical system works by iteration, too, with the same rule applied again and again over very tiny intervals of time. So the x^2 button is a good example of a dynamical system.

Try iterating some other buttons on your calculator. My description here assumes you always start with 0.321, but you can use other starting values if you want. Avoid 0, though. If you press the **cos** (cosine) button about fifty times, you'll find the mysterious number

$$0.739085133,$$

which thereafter just sits there. Again, the iteration of **cos** just settles down to a single value: it converges to a steady state.

The **1/x** button (reciprocal) does something more interesting: the number switches alternately from

$$0.321$$

to

$$3.11526$$

and back again. The iteration is periodic, with period 2. That is, if you hit the button twice, you get back where you started.

The **exp** (exponential) button rapidly blows up to numbers so large that they exceed the calculator's capacity: you get the result 'E' — for 'error'.

Push all the buttons you've got: you'll find that the above outcomes — settle to a steady state, become periodic, or blow up — seem to be the only possible types of behaviour. Considering how many buttons the calculator has, this suggests a certain lack of variety in dynamical systems. But maybe this paucity of behaviour occurs because the buttons on a calculator are designed to do nice things? Nature may not be so accommodating. Let's invent new buttons. What about an x^2-1 button? To simulate it, hit the x^2 button and then $-1 =$. Keep doing this. You soon find you're cycling between 0 and -1, over and over again. The behaviour is just a periodic cycle again. There's nothing more predictable than a periodic universe: watch one cycle, and you know exactly what it will do for ever. Imagine the ease of weather forecasting if a given day of the week always had the same weather!

One last try: a $2x^2 - 1$ button. Now the numbers read

$$0.321, \quad -0.794, \quad 0.261, \quad -0.864, \quad 0.494, \quad -0.513, \quad -0.474, \ldots$$

They jump around a lot, without much pattern...

Aha!

This is chaos.

$2x^2 - 1$ is a simple enough rule. But the results of iterating it don't look so simple: in fact they look random.

Now try the $2x^2 - 1$ button again, but start with 0.322 instead of 0.321. It still looks random — and after just ten iterations it also looks completely different — so different that it's not obvious that the two lists are generated by the same calculator button. Even if you start with 0.3210001, the same thing happens — but now it takes about twenty iterations.

What makes this all the more remarkable is that while $2x^2 - 1$ is so weird, the superficially similar button $x^2 - 1$ is perfectly well behaved.

Why is this? The main reason is that the formula $2x^2 - 1$ magnifies tiny differences in the values of x, so that after a certain number of iterations values that started very close together move pretty much independently. Thus the behaviour is 'unpredictable'. The formula $x^2 - 1$ doesn't do this. It's as simple as that.

Chaos theorists call this phenomenon the butterfly effect. The name comes from a lecture given by the meteorologist Edward Lorenz, although he didn't actually use that phrase. He pointed out that weather is chaotic, and that it is therefore subject to the same sensitivity to tiny changes. If a butterfly flaps its wings today, said Lorenz, then a month later the world's weather will be quite different from what it would have been, had the butterfly *not* flapped its wings.

DON'T BLAME THE BUTTERFLY

Before you rush off to find the Chaos Butterfly and use it to hold the world to ransom, I should point out that in the real world we don't get to run the weather twice, nor do we have only one butterfly to worry about. Lorenz was dramatising the unpredictability of chaos. 'Prediction is very difficult — especially about the future.' So said the Nobel prizewinning physicist Niels Bohr. You don't need to be a Nobel prizewinner to know, from personal experience, that he was absolutely right.

Let's compare two everyday examples of dynamics: weather, and tides. You can find tables of the tides in diaries. But nobody includes a table of the weather. '6 June, full Moon, high tide at 7.42 a.m., sunny periods punctuated by light showers.' No, it doesn't ring true. Weather isn't that predictable.

Yet tides are.

Why are tides predictable, but weather not? Both tides and weather are governed by natural laws. The tides are caused by the gravitational attraction of the Sun and Moon, the weather by the motion of the atmosphere under the influence of heat from the Sun. The law of gravitation is not noticeably simpler than the laws of fluid dynamics, yet for weather the resulting behaviour seems to be far more complicated.

So, if it's not the complexity of the laws, why is there any difference at all? Because the laws for weather generate chaotic dynamics, and those for tides don't. A system can be unpredictable without being random. Laplace notwithstanding, it can be unpredictable even though it is deterministic.

Let's 'compare and contrast'.

A system is random if its future is independent of its past. It has no 'memory', and it is therefore *totally* unpredictable. Previous throws of an unbiased die provide no information about the next throw. Even if you've thrown twenty 6's in a row, the next throw is no more, and no less, likely to be a 6 than it was on every previous throw. There is always one chance in six of throwing a 6.

At the opposite extreme is the 'clockwork universe' behaviour envisaged by Isaac Newton in his laws of motion and gravity. In this view, the entire future — for ever — is completely determined by the present. The clockwork universe ticks for ever, and the only choice available to the deity is the positions of the cogwheels at the moment of creation. Behaviour like this is deterministic, and at first sight 'deterministic' seems to mean the same as 'predictable'.

Chaos sneaks in through a semantic distinction between what is predictable in high philosophical principle, and what you can actually do in practice. Yes, if you know the *exact* state of every particle in the universe now, you can in principle predict the future completely. But in practice you can't have exact information. Chaos occurs when any error, however tiny, grows as it propagates into the future, until eventually its effect becomes so large that the prediction is completely wrong. Chaotic systems are unpredictable in the long term. Just what this means depends on the system: a week for the weather, a microsecond for a turbulent fluid, a hundred million years for the solar system. But whatever the time-horizon may be, your prediction-telescope can't see beyond it.

ATTRACTORS

This description makes chaos seem useless: an obstacle to understanding, not an aid. Even if this were a valid interpretation, the

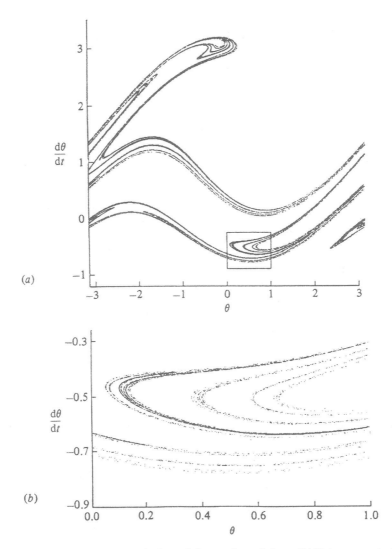

Figure 87 (a) The attractor of a forced damped pendulum. (b) Enlargement of the marked window to show the fractal fine structure. The horizontal coordinate is the angle at which the bob hangs, the vertical coordinate is its angular velocity.

universe wouldn't care very much: chaos is *there*, and it won't go away just because human beings dislike it. But there is a second, far more interesting, side to chaos — hidden patterns. Chaos is behaviour that looks random unless you know how to tease out the secret structure that reflects its deterministic origins. One of the most important secrets is that chaos is confined to well-defined ranges of behaviour, called *attractors*. The motion *on* the attractor is irregular and appears random, but the motions that confine the system *to* that attractor are predictable and regular.

As an analogy, imagine releasing a ping-pong ball on an ocean-covered planet. The ocean's surface is a chaotic place, with waves of all sizes buffeted by winds and pushed to and fro by currents. If you release the ball on the surface of the sea, it will follow a chaotic path. But if you release it from the bottom of the sea, it will float rapidly to the surface. Its upward motion will be relatively simple and predictable, though it will of course respond to some extent to the horizontal currents as well. Similarly, if you release it above the ocean, it will fall through the atmosphere until it reaches the surface. So the motion splits up into two parts. First, a highly regular motion *towards* the attractor (the ocean surface); then a much more irregular motion *on* it.

In this case the attractor is the surface of the entire ocean, which is rather big. It probably doesn't help much in understanding how the ball will move to know that it quickly reaches its attractor. But often the attractor is quite small. Figure 87 shows an attractor for a 'forced damped pendulum'. This is very like the familiar pendulum in a grandfather clock, but with some friction to slow it down and a rhythmic stimulus to keep it going despite that. Unlike the nice, periodic clock pendulum, it moves erratically. It is chaotic. But when you plot out a picture of the states that it takes up, you find a definite, non-fuzzy shape. That's the attractor. It's a very intricate shape, though — whereas the analogous shape for a grandfather clock pendulum is a single closed loop, corresponding to its simple periodicity.

The attractor of a system, with nice, regular dynamics, is very simple — a single point for steady states, a closed loop for periodic ones. A chaotic attractor is utterly different. It has structure on all scales.

It's a fractal.

CHAOS IN THE ASTEROID BELT

For a survey of the applications of chaos, try my book *Does God Play Dice?* To whet your appetite, I'm going to discuss an application to the long-term dynamics of our own dear solar system.

For centuries, astronomers have studied the regularities of the solar system — the orbits of the planets, the phases of the Moon, the paths of comets, the sunspot cycle. Now, equipped with the new mathematics of chaos, they can also study its irregularities. Some of these are important even for the most planetbound of human beings.

We don't normally worry much about the solar system: the Earth has been around for over four billion years, and we expect it to continue to provide a home for us for a good few years yet. So, perhaps, did the dinosaurs — until the notorious K/T meteorite crashed to Earth 65 million years ago and wiped them all out☞. And it's only a short time since Comet Shoemaker-Levy 9 made a spectacular series of impacts on the planet Jupiter, creating shockwaves bigger than the entire Earth. All of which makes our continued existence more fragile than we tend to imagine. The workings of chaos could at any time drop a large asteroid onto the Earth and wreak the kind of havoc described in Larry Niven and Jerry Pournelle's *Lucifer's Hammer* or Gregory Benford's *Shiva Descending*. Moreover, this cosmic destroyer need not be some unsuspected visitor from the depths of interstellar space. It could be any one of thousands of large rocks currently following regular and harmless orbits around the Sun at just the right distance between Mars and Jupiter. At any moment the precise, regular clockwork of the solar system could throw a very unexpected spanner into the machinery of our Earthly existence.

☞ See Walter Alvarez and Frank Asaro, 'An extraterrestrial impact', *Scientific American* (Oct 1990) pp. 44–52. For a contrary view, see Vincent E. Courtillot, 'A volcanic eruption', *Scientific American* (Oct 1990), pp. 53–60.

Paradoxically, it requires a very precise conjunction of circumstances for such a disaster to happen. It all depends on resonances — astronomical phenomena whose periods bear a simple numerical relationship to each other. For example, the Moon always faces the Earth, a 1:1 resonance between its orbital and rotational periods. Mercury takes 87.97 days to revolve once round the Sun, and 58.65 days to rotate once on its axis. Two-thirds of 87.97 is very close to 58.65, so Mercury's orbital and rotational periods are in a 2:3 resonance. Saturn has a large number of satellites, among them Hyperion and Titan. Hyperion takes 21.26 days to complete one orbit, and Titan takes 15.94. The ratio of the two periods is 1.3337, convincingly close to the ratio 4:3.

Resonances are important because they imply that at regular intervals of time — the common period — the bodies in question bear *exactly* the same relationship to each other. This affects their dynamics. Some resonances are stable, and cause the bodies' motions to 'lock together' — just as the Moon's period of rotation on its axis is locked to its period of revolution around the Earth, so that we see only side of the Moon. Others are unstable, and cause wild behaviour. Which of these things happens depends on which resonance you've got.

It is resonances that could at any moment dump the equivalent of a gigatonne hydrogen bomb into our backyard. The effect is related to a long-standing astronomical conundrum, the gaps in the asteroid belt. The largest asteroid, Ceres, is about 940 km across. The smallest are little more than huge rocks, and there are hundreds of thousands of them. Most asteroids circle in the 'main belt' between the orbits of Mars and Jupiter, but a few come much closer to the Sun. The asteroid orbits are not spread uniformly between Mars and Jupiter. Their orbital distances from the Sun tend to cluster around some values and stay away from others. Daniel Kirkwood, an American astronomer who discovered this lack of uniformity in the 1860s, pointed out where the most prominent gaps occur. If a body were to encircle the Sun in one of these 'Kirkwood gaps', then its orbital period would resonate with that of Jupiter. Conclusion: resonance with Jupiter somehow perturbs any bodies in such orbits, and causes some kind of instability which sweeps them away to distances at which resonance no longer occurs. The special role of Jupiter is to be expected — it's so massive in comparison with the other planets.

Until recently, there were no mathematical methods for performing a long-term analysis of any of these resonances, but more powerful computers and new theoretical techniques now exist. The 3:1 resonance, for instance, is pretty well understood. The calculations show that an asteroid, orbiting at a distance that would undergo a 3:1 resonance with Jupiter, can follow a very irregular path. The eccentricity of its orbit — how fat or thin the orbit is — can change violently and almost at random. This is an astronomical example of chaos. The irregularities happen on a time-scale that's short by cosmic standards: about ten thousand years.

A main-belt asteroid whose orbit acquires eccentricity 0.3 or more becomes Mars-crossing, meaning that its orbit can cross that of Mars. Every time it does so, there's a chance that it will come sufficiently close to Mars to be hurled off into some totally different orbit. Until it was realised that chaos could generate high eccentricity, Mars-crossing was not a plausible mechanism for flinging asteroids at the Earth. Asteroids around the 3:1 Kirkwood gap were expected to stay well clear of Mars: there was no reason to expect a sudden change of eccentricity. But now there is such a reason, the mathematics of chaos. The 3:1 Kirkwood gap is there because Mars sweeps it clean, rather than being due to some action of Jupiter. Jupiter creates the resonance that causes the asteroid to become a Mars-crosser, then Mars kicks it away.

Jupiter creates the opening; Mars scores the goal.

Or maybe the touchdown.

Mars might *just* kick the asteroid in our direction (Figure 88). With

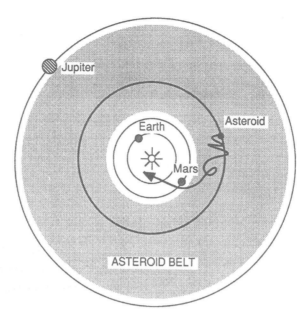

Figure 88   Cosmic soccer played with a chaotic asteroid. Jupiter centres from the corner — Mars scores?

Mars's help, the 3:1 resonance with Jupiter can transport rocks from the asteroid belt into Earth orbit, to burn up as meteorites in our planet's atmosphere.

And, if they're big enough, to burn us up instead.

Maybe it was just such a Martian 'invasion' that killed off the dinosaurs.

It's a sobering thought. Newton's beautiful, regular, clockwork laws are the rules for a random football game, played out by Mars and Jupiter on a cosmic battlefield. This game determines whether or not life continues to survive on Earth. Newton thought he'd ruled out divine intervention, but the unruly gods of the Enuma Elish are alive — and kicking. The chance that this celestial version of Russian roulette will prove fatal is admittedly very, very tiny — but it's there. The solar system is — literally — dicing with death.

Ours.

NOW FOR THE HAPPY ENDING . . .

Since I made such a big fuss about chaos not being a Cassandra, I can hardly stop there, however dramatic an exit the total annihilation of humanity might offer.

Chaos has a good side, too.

For example — and this is just one of many — one consequence of the butterfly effect is that you can make big changes to a chaotic system with minimal effort. You can control it, easily. It sounds paradoxical, for chaotic dynamics looks completely out of control. But chaos has hidden patterns, and it is those patterns that make it controllable. An amazing new technique, known as chaotic control, has grown from this insight: the person most responsible is Jim Yorke, although there are several others.

One potential application is to build an intelligent heart pacemaker. It has been known for some years that certain kinds of heart disease involve the onset of chaotic dynamics. A tiny computer chip could be programmed to pick up these irregularities, and switch on a 'chaotic control' system that stabilises the heart back to more desirable regular rhythms. Current pacemakers do this with a big electrical current; an intelligent pacemaker would achieve better results with much less power, so its power supply would last far longer, avoiding the need to operate on the patient to put in a new 'battery'.

The same technique might be able to make passenger aircraft more efficient by getting rid of turbulence in the airflow over the wings. Thousands of tiny sensors would monitor the flow, and a computer system would exploit the chaotic behaviour of turbulence to manipulate thousands of tiny flaps, many times every second, cancelling the turbulence out again.

Even the space programme could benefit. Until recently, it was generally assumed that the most efficient orbit to send a package from the Earth to Moon — or from any world to any other — is a 'Hohmann ellipse'. This is an ellipse that surrounds both worlds, coming close to one of them at each end. However, it has now been discovered that a much more complicated orbit — one that is practically feasible only because of chaotic control methods — can do the same job using a lot less fuel. It takes longer — up to two years — so it wouldn't be used to send an astronaut.

But it could be used to supply a lunar colony.

Once more, the magic of the maze astounds us. When mathematicians first began to investigate chaos, they did so out of sheer curiosity. For years they were told that chaos was just a fad, a fashion, a load of media hype with no intellectual content and no uses whatsoever. Suddenly all of this turns out to be nonsense: the bug of fashion has bitten the ankles of its detractors, the potential applications of chaos seem inexhaustible.

When you explore the magical maze with an imaginative mind, marvels await you.

Index